Advances in Geographical and Environmental Sciences

Series Editor

R. B. Singh, University of Delhi, Delhi, India

Advances in Geographical and Environmental Sciences synthesizes series diagnostigation and prognostication of earth environment, incorporating challenging interactive areas within ecological envelope of geosphere, biosphere, hydrosphere, atmosphere and cryosphere. It deals with land use land cover change (LUCC), urbanization, energy flux, land-ocean fluxes, climate, food security, ecohydrology, biodiversity, natural hazards and disasters, human health and their mutual interaction and feedback mechanism in order to contribute towards sustainable future. The geosciences methods range from traditional field techniques and conventional data collection, use of remote sensing and geographical information system, computer aided technique to advance geostatistical and dynamic modeling.

The series integrate past, present and future of geospheric attributes incorporating biophysical and human dimensions in spatio-temporal perspectives. The geosciences, encompassing land-ocean-atmosphere interaction is considered as a vital component in the context of environmental issues, especially in observation and prediction of air and water pollution, global warming and urban heat islands. It is important to communicate the advances in geosciences to increase resilience of society through capacity building for mitigating the impact of natural hazards and disasters. Sustainability of human society depends strongly on the earth environment, and thus the development of geosciences is critical for a better understanding of our living environment, and its sustainable development.

Geoscience also has the responsibility to not confine itself to addressing current problems but it is also developing a framework to address future issues. In order to build a 'Future Earth Model' for understanding and predicting the functioning of the whole climatic system, collaboration of experts in the traditional earth disciplines as well as in ecology, information technology, instrumentation and complex system is essential, through initiatives from human geoscientists. Thus human geosceince is emerging as key policy science for contributing towards sustainability/survivality science together with future earth initiative.

Advances in Geographical and Environmental Sciences series publishes books that contain novel approaches in tackling issues of human geoscience in its broadest sense—books in the series should focus on true progress in a particular area or region. The series includes monographs and edited volumes without any limitations in the page numbers.

Utsav Kumar Singh

Human Development in Bihar, India

Experiences of Millennium Development Goals

Springer

Utsav Kumar Singh
Department of Economics
Shaheed Bhagat Singh College
University of Delhi
New Delhi, India

ISSN 2198-3542 ISSN 2198-3550 (electronic)
Advances in Geographical and Environmental Sciences
ISBN 978-981-19-3623-4 ISBN 978-981-19-3624-1 (eBook)
https://doi.org/10.1007/978-981-19-3624-1

This Springer imprint is published by the registered company Springer Nature Singapore Pte Ltd.
The registered company address is: 152 Beach Road, #21-01/04 Gateway East, Singapore 189721, Singapore

To Prof. R. B. Singh, among the rarest who is an inspiration both in his professional career and personal life.

Foreword

Economics studies the characteristics of economic space and society. Economists identify and analyze the spatial patterns of human–economy relations that shape our lives and society. I am pleased that Dr. Utsav Kumar Singh is bringing out a book on human development in the framework of post-MDGs agenda for development in India. The Millennium Summit of September 2000 adopted the UN Millennium Declaration for a new global partnership to reduce poverty under the framework of Millennium Development Goals (MDG). The MDGs are a set of numerical and time-bound targets related to key achievements in human development. They include halving poverty alleviation and hunger, achieving universal primary education and gender equality, reducing infant and child mortality by two-thirds and maternal mortality by three–quarters, reversing the spread of HIV/AIDS and other communicable diseases, and halving the proportion of people without access to safe water. The mandate says that the responsibilities for achieving these goals lie with the individual countries.

This book highlights the progress being made in many areas of MDGs targets. Since the implementation of MDGs, Bihar is rising from the dust-biting state to the fastest growing economy in India. In the past two decades with its various policies in accordance with the central program in line with MDGs, the state reduced the proportion of people living below the poverty line. People are getting better health facilities, the State has achieved the target of universal enrollment ratio and gender enrolment ratio, maternal and infant mortality have reduced, and reservation in Panchayati Raj Institution has empowered the women further. More households have got connected through the clean and safe drinking water supply. Open defecation has been declining. Internet and mobile connectivity have improved. Connectivity has become cheaper, affordable, and geographical penetration due to expanding network infrastructure. This book aimed at utilizing the available data from reliable sources in assessing the progress at the national and subnational levels. The necessary ingredients for human development have been incorporated into the growth story of Bihar in the past two decades, however, there is still a lot left to be done. Undoubtedly, the state has gained pace in removing the obstacles to human development, nonetheless a slow one, and in the future it will capitalize on these preliminary successes to come out as a shining star. The chapters include Introduction, Implementation of MDGs in Bihar, Parameters of MDG, Post MDGs-Target for SDGs, Conclusion, etc. The book is extremely useful and provides new insights for policy-makers especially for the State of Bihar and is indispensable for any organization working for improving the life chances of the marginalized and deprived section. The book is also a very important source of knowledge for students, teachers, and researchers from Economics, Geography, and Development Studies.

August, 2019

Prof. R. B. Singh
Secretary General and Treasurer-IGU
Chair-Research Council: CSIR-CFTRI
Government of India
Mysore, India
rbsgeo2@hotmail.com

Professor and Head
Department of Geography
Delhi School of Economics
University of Delhi
New Delhi, India

Preface

Being born and brought up in Bihar, there has always been a personal and emotional connection with the state. I moved to Delhi for my higher education but have always tried to be well-read and aware of the socioeconomic and political problems of Bihar. A major issue that struck me very early on was the lack of scholarly attention on the issues concerning Bihar in the national mainstream media. So, I strove to equip myself with the technical skills of research in which I benefitted immensely from my Ph.D. coursework. I learned to conduct closer observation of the focus groups, developing a 'people's perspective' about the issues concerned, and interpreting the role of other simultaneous factors which might shed altogether new light on the research concerned. I made Bihar my area of focus, with frequent academic visits during my Ph.D. studies. These visits helped me develop an idea about the daily toil of the common people at the grassroots level. Bihar falls in the category of poor performer states, which makes the ordinary citizen very vulnerable and limits them to aspire for lowly incentives instead of curating the possibilities in the future. While researching Millennium Development Goals (MDGs) I realized that there was a lack of serious research on the topic, especially with respect to assessing those goals in the case of Bihar which motivated me to put forth this academic work.

I see Bihar against its history as the land of the great Nalanda University, the place where the first political map of India, i.e., "Akhand Bharat" was consolidated and carved. The glorious past of Bihar is reflected in the Mauryan and the Gupta empires, the periods in Indian history that are accorded the "golden" status. It is a credit to these kingdoms that history records free medical facilities that were offered within the kingdom. The sophisticated level of statecraft of this period is also evident in the Arthashastra, a treatise on the political economy by Kautilya. In one part of the treatise, Kautilya writes, "*By maintaining order, the king can preserve what he already has, acquire new possessions, augment his wealth and power and share the benefits of improvement with those worthy of such gifts. The progress of this world depends on the maintenance of order and the proper functioning of government*". Today, even as India marches toward robust economic growth, the benefits of progress attained have not benefited the common people on the lower rungs of the ladder. This holds true, particularly for the state of Bihar, which continues to struggle even in terms of providing basic amenities. Bihar has done well during the period of Millennium Development Goals (2000–2015), achieving considerable success in the goals of hunger, poverty, and gender but performed poorly in health. The onset of the COVID-19 pandemic exposed the crumbling social infrastructure and all the claims of good governance. The devastating impact of the COVID-19 pandemic was felt in every sector; the heart-wrenching images of the migrating laborers, the poor medical infrastructure crumbling under pressure, the drying up of the income sources of the people, etc.

To finish the unfinished goals of MDGs, world leaders adopted the Sustainable Development Goals (SDGs) on January 1, 2016, for the next 15 years. The UNDP has provided a set of targets in form of Sustainable Development Goals (SDGs) to be achieved by 2030. SDGs—the set of 17 goals and 169 targets—pledges the vision, "no one is left behind". Regrettably, the world came to a standstill amidst the COVID-19 pandemic which halted the progress toward such objectives.

India is a union of unbalanced states. Even with all its initiatives as a nation, India consists of unequal states, which do not conform to the breakthrough claimed by the nation. The state of Bihar is an apparent example. Bihar was the worst performer in the second edition of SDGs Index based on goals and targets in NITI Aayog's report-2019. However, despite its misfortunes, Bihar is still one of the youngest states in India with approximately 65% of its population in the working-age group; therefore, the realization of the demographic dividends has the potential to help the state to achieve these targets better.

This book is an attempt to shed light on the issues of the contemporary state of Bihar, on the lines of its performance in the Millennium Development Goals (MDGs) indicators and the initiatives taken by the Central and State government toward realizing the Sustainable Development Goals (SDGs). At the same time, this book analyses the impact of COVID-19 on India in general and on Bihar in particular, concluding with suggestive measures.

New Delhi, India Dr. Utsav Kumar Singh

Acknowledgments

It would not be possible to repay the debt of gratitude that I have incurred in writing this book. To all who have been a part of my journey and to all who shared my burden, I owe them my sincerest gratitude.

I would like to acknowledge the guidance and encouragement that I received from my mentor Late Prof. R. B. Singh, Treasurer and Secretary-General, International Geography Union. He was also the former Head of the Department of Geography, University of Delhi, and a man of powerful intellect who has had an immense influence in formulating my research ideas.

I convey sincere gratitude to Prof. Rekha Saxena, Department of Political Science, University of Delhi for introducing new and critical perspectives to my postdoctoral research. Her piercing insights and meticulous approach to research have been a constant guide in the progress of my work.

I would like to thank my Ph.D. supervisor Prof. Suresh Kumar, Department of African Studies, University of Delhi for his guidance and valuable feedback.

I am thankful to Ms. Niti Rawat, Research Assistant in Health Education and Livelihood Promoter (HELP), for her help in coordinating and formulating my Research work.

I would like to thank Thejalhoukho Casavi for his contribution in formalizing this research. I would also like to thank Mr. Deepak Kumar Pathak, Department of Economics, Delhi School of Economics, University of Delhi for his valuable assistance in the collection and analysis of data from various secondary sources.

Gratitude would be incomplete without mentioning my wife Ms. Rachana Tomar, Research Faculty, University of Vanderbilt, Tennessee, whose steadfast love and unshaken belief in my research work have been my biggest motivation. I also thank my parents, not only for their love and support but also for helping me to see Bihar from many different perspectives.

Last but not least, I would like to thank my hostel (Gwyer Hall) and Department Colleagues for participating in lively discussions over tea and coffee. These discussions broadened my research and helped me gain an interdisciplinary approach toward various issues of Development for the state of Bihar as well as our country.

I regret the inability to mention each and every person whose help has been indispensable for the completion of this work. To you all, I offer my deepest gratitude.

Dr. Utsav Kumar Singh

Contents

1 Introduction ... 1
References ... 7

2 Introspection of Human Development in India ... 9
2.1 Introduction ... 9
2.2 Literature Survey ... 11
2.3 Millennium Development Goals (MDGs) ... 14
2.4 Goals and Performance ... 15
2.4.1 Goal 1: Eradicate Extreme Poverty and Hunger ... 15
2.4.2 MDG2 Achieve Universal Primary Education ... 16
2.4.3 MDG3 Gender Parity and Women Empowerment ... 17
2.4.4 Goal 4: Reduce Child Mortality ... 20
2.4.5 Goal 5: Improve Maternal Health ... 22
2.4.6 Goal 6: Reverse the Spread of Communicable Diseases (HIV/AIDS, Malaria and Other Diseases) ... 24
2.4.7 Goal 7: Ensure Environment Sustainability ... 27
2.4.8 Goal 8: Fostering Partnership for Development ... 30
2.5 MDGs Implementation ... 31
2.6 MDGs Challenges ... 32
2.7 Conclusion ... 33
References ... 35

3 Appraisal of Human Development Through the Lens of Millennium Development Goals in Bihar ... 39
3.1 Introduction ... 39
3.2 Growth Process and Development ... 40
3.3 Program and Policies Framed to Achieve MDG in Bihar ... 47
3.4 Convergence of Millennium Development Golas (MDGs) and Public Policies in Bihar ... 49
3.4.1 MDG1 ... 49
3.5 Poverty Measures in India ... 50
3.6 Pro-poor Policy in Bihar ... 53

3.6.1 The Mahatma Gandhi National Rural Employment Guarantee Scheme (MNREGA) 54
3.6.2 Deen Dayal Antoyoday Yojana (DAY-NRLM) 56
3.6.3 Jeevika 56
3.6.4 MDG2 57
3.6.5 Education as Fundamental Right 58
3.6.6 Right to Education Act 2009 59
3.6.7 Performance of Goal 60
3.6.8 MDG3 61
3.6.9 National Program for Education of Girls at Elementary Level (NPEGEL) 65
3.7 Discussion 82
3.8 MDG's Challenges in Bihar 84
3.9 Conclusion 85
References 85

4 Quantitative Analysis of Human Development in India 91
4.1 Introduction 91
4.2 Human Development Index (HDI) 91
4.3 HDI in India 94
4.4 Methodology 95
4.4.1 Coefficient of Determination 96
4.4.2 Time Trend 96
4.5 Results 96
References 97

5 Sustaining Development Process: Set of New Targets 99
5.1 Introduction 99
5.1.1 Sustainable Development Goals 100
5.2 Transition of Goals (From MDGs to SDGs) 101
5.3 SDGs Baseline 102
5.3.1 Synergy Between SDGs 112
5.4 Implementation of SDG in India 113
5.5 Convergence of SDGs with India's Priority 113
5.6 Social Goals 114
5.7 Economic Goals 115
5.8 Environmental Goals 115
5.9 Peace Justice, Governance, and Partnership 115
5.10 Discussion 116
5.10.1 Social Safety Net 116
5.10.2 Skill Education 118
5.10.3 Health 120
5.10.4 Challenges 120
5.11 Conclusion 122
References 122

6 Impact of COVID-19 on Sustainable Development Goals 131
6.1 Introduction 131
6.2 Poverty 132
6.3 Impact of Pandemic on Goals 134
6.4 Conclusion 140
References 140

7 Conclusion and Way Forward 149
7.1 Introduction 149
7.2 Achievement During MDGs 150
7.3 Conclusion 154
7.4 The Way Ahead 159
References 161

Abbreviations

ABET	Adult Basic Education and Training
AET	Adult Education and Training
AIBP	Accelerated Irrigation Benefit Program
AMRUT	Atal Mission for Rejuvenation and Urban Transformation
ANA	Annual National Assessment
ANM	Auxiliary Nurse cum-Midwife
AREP	Accelerated Rural Electrification Program
ASER	Annual Survey of Education Report
ASHA	Accredited Social Health Activities
ASI	Annual Survey of Industries
ASIDI	Accelerated Schools Infrastructure Delivery Initiative
AWC	Anganwadi Centers
AWW	Anganwadi Worker
BED	Bachelor of Education
BERC	Bihar Electricity Regulatory Commission
BIADA	Bihar Industrial Area Development Authority
BIGWIS	Bihar Ground Water Irrigation Scheme
BREDA	Bihar Renewable Energy Development Agency
BRGF	Backward Region Grant Fund
BSDM	Bihar Skill Development Mission
BSHP	Bihar State Highway Program
BSHPC	Bihar State Hydro-Electric Power Corporation
BSUP	Basic Services for Urban Program
BSWAN	Bihar State Wide Area Network
BUDCL	Bihar Urban Development Corporation Limited
BUIDCO	Bihar Urban Infrastructure Development Corporation Limited
CADA	Command Area Development Agencies
CADWM	Command Area Development and Water Management
CAGR	Compounded Annual Growth Rate
CBS	Core Banking Services
CLRI	Central Leather Research Institute

COMFED	Cooperative Milk Producers Federation Limited
CPTD	Continuous Professional Teacher Development
CPWD	Central Public Works Department
CSO	Central Statistical Organisation
CSS	Centrally Sponsored Scheme
CST	Central Sales Tax
DBE	Department of Basic Education
DCRF	Debt Consolidation and Relief Facility
DDUGJY	Dee Dayal Upadhyay's Gram Jyoti Yojana
DFID	Department for International Development
DHET	Department of Higher Education and Training (Established in 2009)
DIC	District Industries Centre
DoE	Department of Education
DoL	Department of Labour
DPAP	Drought Prone Area Program
ECD	Early Childhood Development
ERC	Education Resource Centres
FET	Further Education and Training
FRBMA	Fiscal Responsibility and Budget Management Act
GDDP	Gross District Domestic Product
GDP	Gross Domestic Product
GET	General Education and Training
GHS	General Household Survey
HE	Higher Education
HSRC	Human Sciences Research Council
IAY	Indira Awas Yojana
ICDS	Integrated Child Development Scheme
IGMSY	Indira Gandhi Matriva Sahyog Yojana
IHSDP	Integrated Housing and Slum Development Program
ILO	International Labour Organization
KGBV	Kasturba Gandhi Balika Vidyalaya
LER	Learner-to-educator ratio
MDG	Millennium Development Goal
MGNREGA	Mahatma Gandhi National Rural Employment Guarantee Act
MMGSY	Mukhyamantri Gram Sadak Yojana
MNSY	Mukhyamantri Nari Shakti Yojana
MSME	Micro, Small and Medium Enterprises
NABARD	National Bank of Agricultural and Rural Development
NASCA	National Senior Certificate for Adults
NDMA	National Disaster Management Authority
NDP	National Development Plan
NDRF	National Disaster Response Force
NEEDU	National Education Evaluation and Development Unit
NEET	Young people "Not in Employment, Education or Training"

NEIMS	National Education Infrastructure Management System
NHAI	National Highway Authority of India
NQF	National Qualifications Framework
NRHM	National Rural Health Mission
NRLM	National rural Livelihood Mission
NSC	National Senior Certificate
NSF	National Skills Fund
NSSO	National Sample Survey Organisation
NULM	National Urban Livelihood mission
NW	North West
OECD	Organization for Economic Co-operation and Development
PACS	Primary Agricultural Cooperative Societies
PALC	Public Adult Learning Centre
PDS	Public Distribution System
PHC	Primary Health Centre
PMEGP	Prime Minister Employment Generation Program
PPN	Post-provisioning Norm
PRI	Panchayat I Raj Institution
PSE	Post-school Education System
PSU	Public Sector Undertakings
R-APDRP	Restructured Accelerated Power Development and Reforms Program
RAY	Rajiv Awas Yojana
RGSEAG	Rajiv Gandhi Scheme for Empowerment of Adolescent Girls
RIDF	Rural Infrastructure Development Fund
RSETI	Rural Self Employment Training Institute
RSVY	Rashtriya Sam Vikas Yojana
SCADA	Supervisory Control and Data Acquisition
SDG	Sustainable Development Goals
SDRF	State Disaster Response Force
SETA	Sector Education and Training Authority
SIPB	State investment Promotion Board
SJVNL	Sutluj Jal Vidyut Nigam Limited
SLBC	State Level Bankers Committee
SPUR	Support program for Urban Reforms
SRI	System of Rice Intensification
SRN	School Register of Needs
SRR	Seed Replacement Rate
SRS	Sample Registration System
TDI	Teacher Development Institutes
TVET	Technical and Vocational Education and Training
UIDSSMT	Urban Infrastructure Development Scheme for Small and Medium Towns

UIG	Urban Infrastructure and Governance
UN	United Nations
UNDP	United Nations Development Program
UNESCO	United Nations Educational, Scientific and Cultural Organization
WHO	World Health Organization
WUA	Water Users Association

List of Figures

Fig. 2.1	Poverty head count ratio, India	16
Fig. 2.2	Proportion of Female in PLFS, India	18
Fig. 2.3	Immunization against measles in India	21
Fig. 2.4	Trends of HIV incidence in India	22
Fig. 2.5	Household's having access to improved sources of drinking water	26
Fig. 2.6	Household's having bathroom and latrine, India	30
Fig. 2.7	Telephone user/100 persons, India	31
Fig. 3.1	Per capita state domestic product of Bihar and NNI (at constant price) (INR)	43
Fig. 3.2	Per capita expenditure on social capital in Bihar and All States of India	44
Fig. 3.3	Poverty Head Count Ratio, Bihar	49
Fig. 3.4	Gross enrolment ratio under Primary School, Bihar	57
Fig. 3.5	Enrolment in primary classes (Bihar)	60
Fig. 3.6	Students–classroom ratio (Bihar)	61
Fig. 3.7	Gender parity index, primary and tertiary level	63
Fig. 3.8	Women representation in Sixteenth Lok Sabha and share of Bihar	64
Fig. 3.9	Trends of malnutrition (Bihar)	65
Fig. 3.10	Proportion of under-weight children below 3 years of age	66
Fig. 3.11	Trends of Under-five mortality rate, Bihar	69
Fig. 3.12	Districts trends of infant mortality and neonatal mortality, Bihar	69
Fig. 3.13	Immunization against Measles, Bihar	70
Fig. 3.14	Evidences of maternal mortality rate India and Bihar	73
Fig. 3.15	Distribution of Forest area, Bihar	79
Fig. 3.16	Forestation trends of Bihar and India	79

Fig. 3.17 Number of telephone connections, Bihar 82
Fig. 7.1 Trend of PHCR of India and Bihar 150
Fig. 7.2 Trend of women participation in Indian Parliament 151

List of Tables

Table 2.1	MDG1 (targets and indicators)	15
Table 2.2	Population living below $1.25/day (India and Bihar) and rate of change to meet the target	16
Table 2.3	MDG2: targets and indicators	16
Table 2.4	Net primary enrolment (India)	17
Table 2.5	MDG3 targets and indicators	17
Table 2.6	Proportion of women in India's National Parliament	19
Table 2.7	MDG4 reduce child mortality	20
Table 2.8	MDG5, targets and indicators	23
Table 2.9	MDG6 targets and indicators	25
Table 2.10	Status of HIV/AIDS in India	25
Table 2.11	Incidence of tuberculosis in India	25
Table 2.12	MDG7 targets and indicators	28
Table 2.13	India's forest cover	28
Table 2.14	MDG8 targets and indicators	31
Table 3.1	Facts and figures of Bihar and India	40
Table 3.2	Socio-economic and caste census	41
Table 3.3	Health infrastructure of Bihar	45
Table 3.4	Social welfare scheme of union government and government of Bihar	48
Table 3.5	Poverty incidence, Bihar and India	51
Table 3.6	Distribution of population as per their earning	52
Table 3.7	Job provided under MGNREGA in the past 10 years	55
Table 3.8	Literacy rate, Bihar	62
Table 3.9	Trends of universal coverage in measles immunization of one-year-old children	71
Table 3.10	Distribution of ASHA workers in Bihar	74
Table 3.11	Number of health personnel, Bihar	75
Table 3.12	Incidence of Institutional deliveries, Bihar	76
Table 3.13	Prevalence of major disease, Bihar	77
Table 3.14	Demographic trends of Bihar	83

Table 4.1 HDI ranking, India 95

Table 5.1 Transition of goals (from MDGs to SDGs) 102

Table 5.2 National indicator framework (NIF) 114

Table 7.1 Proportion of 1-year-old children immunized against measles percentage in India 152

Chapter 1
Introduction

Abstract This chapter provides a brief introduction to human development in Bihar. There are various approaches to examining the development process of Bihar. One practical way is to juxtapose the state policies for human development and the targets set under the Millennium Development Goals. Against this, we can further explore the shortcoming and challenges to meeting the Sustainable development Goals.

Keywords Bihar · India · Development · MDG · SDG

The state of Bihar is the least urbanized state in India. The urban population is about 10% of the total population of the state while the population employed in the agricultural sector forms 76% of its workforce (Government of Bihar 2022). The per capita agricultural income in Bihar is half that of India. The state paints a gloomy picture of its underperformance in terms of generating employment in nonagricultural sectors when compared to the other states in India. The rural industries like cotton, jute, etc. have witnessed a downfall. The few changes observed of late in Bihar have been the result of the attention it received in the past 4–5 years, which comes as a breakthrough for a state which has been the subject of perpetual backwardness.

Bihar as a political entity, as a kingdom, and as a state within the Republic of India, has its own distinct identity (Thapar 1966). As a noted historian puts it, the history of ancient India is the history of ancient Bihar. The achievements in the areas like education, governance, and polity, carved a niche for India in the respective areas. Although the state of Bihar was celebrated for its role in trade and economic engagement, the present scenario provides little cause for celebration.

The levels of output of Bihar have been the smallest among all the major states of India. Even in terms of the Human Development Index, the performance of Bihar in realms like access to infrastructure, healthcare, education, law and order, etc. depict a poor picture for the state. It seems that the performance of Bihar when compared to India creates a huge wedge. The gap between the state of Bihar and India's achievement is so large that from the mid-1980s, the status of Bihar was deemed as a "Basket-Case" with little expectation of growth, for the latter half of the twentieth century.

U. K. Singh, *Human Development in Bihar, India*, Advances in Geographical and Environmental Sciences, https://doi.org/10.1007/978-981-19-3624-1_1

The data available on the state income for Bihar indicates steady growth for the past 6 years. During the first 5 years post its separation from Jharkhand in 2000, the annual growth rate was only 4.42% when calculated at constant prices. This only created an abysmal picture for the already underperforming economy of Bihar after the separation. The reason was mainly due to the asymmetric distribution of resources between the state of Bihar and Jharkhand. However, the policies pursued by the government resulted in an optimistic growth rate of 11.36% annually in the period 2004–15 to 2010–11 through its investment patterns. The average annual plan size increased from Rs 1200 crores (Tenth Plan, 2002–07) to 15,000 crores (Eleventh Plan, 2007–12). Further, there has been a promising investment portfolio with its focus on the areas like social delivery of services and infrastructural development. As claimed by the experts, the figures only signify the "take off" stage—a sustained development path.

The chronology of Bihar's development can be divided into three separate time periods, the sub-periods being 1980–2000, 2000–2005, and post-2005 on an analytical basis. The First period (1980–2000) is the pre-separation period of Bihar; November 2000 brought about a shift in terms of its past existence. The districts that constitute Bihar and Jharkhand today are different both socially and economically; for example, while the population of Jharkhand is largely tribal with limited caste identity, caste has historically been the main factor of polarization and exclusion in Bihar (Sharma 2005). The post-bifurcation period (2000–2005) reveals different social and economic dynamics as the manufacturing units and other power generation units went away with the state of Jharkhand. The share of the agriculture sector increased from 36.5 to 40.4% while that of the industry decreased from 22.5 to 4.6% of the NSDP. However, there was an increase in the services sector from 36 to 50% within a year. The loss of industry led to a sharp drop in the non-tax revenue from the same sector (Mukherji and Mukherji September 2012). The decline in the fiscal space was accrued to the bifurcation of the states. In terms of economic development, 2000–05 were trying years. The third period extends after 2005; this emerged as a period of holistic improvement with major changes in policies, administrative and overall changes, and rapid economic growth.

There exists a wide range of intra-state disparities, which emerges as a challenge to the development of Bihar. The state of Bihar is far behind the other states in terms of per capita income. The National Sample Survey for 2004–05 gives the incidence of poverty as 42.1% for rural Bihar which is even higher than 28.3% for rural India. The social and economic backwardness of the state of Bihar dates back to colonial times. The semi-feudalistic structure in Bihar originated in colonial times when the upper castes held the reins of economic as well as political power. The post-independence period witnessed a surge in the upper-middle castes for both the economic as well as political ascendancy alongside military mobilization of the poor peasantry of the low castes. These developments led to a withdrawal from the hitherto caste-based hierarchy for the rural areas, yet it did not end the divisions in landholdings. The economic growth in Bihar was centered around the sectors like construction, restaurants, hotels, communication, and trade. These areas are nonagricultural in nature, yet approximately 81% of the rural workforce is employed in agriculture for a state which has 90% of the population residing in rural areas (NITI Ayog, 2008). The social inequalities in the state of Bihar emanate from caste, religion, and gender, which are clearly reflected in the areas like health and education indicators. There is uneven distribution of the benefits from public investment and development endeavors at the village level clearly bring out these limitations. Further, the initiatives undertaken only underpin the existing socioeconomic inequalities in the rural areas. There is inadequate attention given to policies and programs such as poverty eradication and development. The virtues like accountability, efficiency, and transparency are a dire need for the development of the state.

The state of Bihar is also the poorest state in India. The agriculture sector which employs the great majority of the population of the state has always been the gamble of monsoons, a trend that is prevalent in the country as a whole. This vulnerability of agriculture to the weather poses a major challenge for food security, economic development, household welfare, etc. The smallholders and agricultural laborers who have very few assets and limited access to credit and insurance bear the brunt. Frequent occurrence of droughts adversely impacts agricultural production reducing the output of agriculture. Kharif paddy, the crop with the highest fraction of sown area in the state, remains most affected. Bihar is the most rural state of India and it has the highest working population working in agriculture. There is a high level of dependence on agriculture in Bihar, which means that drought-induced recession in the agrarian economy not only adversely impacts the poverty levels but also the consumption levels of the household.

The social safety nets program is implemented by the government of India and nearly 3% of the country's annual budget is spent on such programs. These programs which are meant for the poor play a significant role and emerge as their savior in times of droughts. The times of droughts are characterized by not only deprivation of job opportunities for the poor but also food unavailability. These programs play an essential role in extending relief for the drought-affected districts in Bihar. However, the shortcomings of these programs are also many as many deserving farmers are not aware of the subsidy schemes granted by the government. There are many farmers who are not aware of such government benefits and others who are baffled by the high transaction costs to avail of these schemes. Further, these schemes also tend to have delayed the delivery of the subsidies or payments. This renders the schemes less useful for the small farmers who are not only risk-averse but also liquidity constrained. The subsidy payments are made only after it is officially declared that drought has hit the area from the district headquarters. Once the drought is declared, the amount is sent to block development officers, who transfer the amount to the mukhiyas of the respective panchayats and when the land revenue receipts and the receipts for the purchase of the diesel are provided the mukhiya is required to disburse the subsidy and also submit the details about the utilization of the subsidies. The delays in each and every step of the disbursement process only make the entire process unconducive for the poor farmers. Thus, fewer farmers apply for seeking the benefits of the subsidy scheme and all this makes the provision of the subsidy only less effective in terms of its reach and depth.

The droughts in the past have varied in their intensity from moderate to severe. The years from 2009 to 2013 have witnessed frequent droughts. Five out of the 38 districts in the state (Khagaria, Pashchim, Champaran, Saharsa, Siwan, and Sheohar) experienced drought in all five years from 2009 to 2013, 14 districts experienced droughts in 4 out of 5 years, 14 districts witnessed 3 years of droughts and the remaining 5 saw 2 years of drought. There is a dire need for initiatives to adapt agriculture to the changing climate. Further, effective social safety net programs can help to prevent the negative shocks of weather on agricultural production.

The current millennium has introduced a multitude of challenges. These challenges vary from poverty, diseases, rising population, malnourishment, lack of education, environmental degradation, gender inequality, etc. The Millennium Development Goals were conceived to tackle these impediments. The MDGs covered these issues in a set of 8 goals, 21 targets, and 43 indicators. The MDGs work toward the achievement of their goals by the generation of incentives, in particular the financial incentives to lay a foundation for a better future.

The MDGs originated from the United Nations Millennium Declaration adopted on 8th September 2000 in the General Assembly. Their focus varied from peace, disarmament, development, poverty eradication, environment protection, human rights, democracy, good governance, protecting the vulnerable sections, etc. However, the progress of these goals as measured by the indicators varies across the regions and the states. The world anticipated that by the year 2015, most of these goals would have been achieved, yet there exist areas where little progress has been made and the state of Bihar has been an underperformer as compared to the other states of India. The following sections of this book shall compare and contrast the performance of the state of Bihar in the Millennium Development Goals like eradication of poverty and hunger, maternal health, gender parity, and women empowerment, reduction in child mortality, prevention against HIV/AIDS, malaria and other diseases, environment sustainability, creation of a global partnership for development. As the world prepares itself for the next milestone with the Sustainable Development Goals and embarks on the journey to its fulfillment by 2030. The book draws light on the achievements of the MDGs in Bihar and further brings out the challenges to the Sustainable Development Goals.

The Government of Bihar has emphasized areas like Personal Development, Human development, social development, economic development, and the environment besides the provision of basic amenities, social security, and good governance for the rural areas. While areas like human, social and economic development are critical for the development of the state growth and development, good and just governance is indispensable to successfully implement these schemes and policies. According to the report Samanvay (A compilation of Central sector, Centrally sponsored, and state schemes for convergence under the Saansad Adarsh Gram Yojana), human development initiatives undertaken include Universal access to Basic health and facilities consisting of health care, medical examination, total immunization, initiatives to balance the sex ratio, 100% institutional delivery, initiatives to improve the nutrition status for all, with special focus on children, adolescent girls, pregnant women, and lactating mothers, etc. Further, there are schemes that have a special focus on the special needs of Persons with disabilities (PWDs), especially children and women. Education is a basic building block for human development and hence it is an area that cannot be compromised, there are schemes that provide Universal access to educational facilities up to class X and retention besides the schemes with Adult Literacy and E-literacy as a focal point.

The basic amenities are being emphasized through a variety of measures such as pucca houses for all houseless poor/ poor living in kutcha houses, land rights and land records transparency, separate activities for water use for drinking purposes, treated piped water, all-weather roads with drainage connectivity, electric supply to all households and street lights with surety of alternative source of power supply, particularly solar energy; *Pucca* (commented) infrastructure for the public, health and social infrastructure such as schools, hospitals, gram panchayat office and libraries, crematorium, etc.; provision for village infrastructure including PDS outlets, kiosks of micro bank/post office/ATM, tele-connectivity, etc.

Personal Development entails measures like hygiene behavior and practices by constructing specific sanitary lavatories dedicated to the households living below poverty lines, converting dry latrines into economical sanitary latrines. At the request of panchayats in land-scarce villages, where there are no spaces for sanitation purposes, villages are provided with sanitary complexes dedicated to women with the service of hand pumps for clean drinking water, bathing, and washing to maintain proper hygiene and sanitation. In such cases, the responsibility of maintenance of a dedicated women's complex is willingly shared by panchayat committees. Further, for total sanitation coverage, a sanitary network is established through a chain of drains and soak pits; dedicated waste management for solid waste and liquid waste. There is a need to educate people through dedicated health and hygiene awareness campaigns.

Social Security policy measures include pensions for all eligible families—old age, disability and old age, disability and widow, insurance schemes, and PDS-universal access to all eligible households. The Social Development policies and schemes pivot around the activities for the promotion of voluntarism like Bharat Nirman Volunteers, the capacity building where the people can fully contribute and participate in the local development, activities for the violence and crime-free village, and activities that honor the village elders, local role models especially women, freedom fighters and martyrs, celebrating the folk arts and festivals. As for economic development, there are activities that promote diversified agricultural and allied livelihoods, including livestock and horticulture, rural industrialization, skill development of all eligible youth for self-employment and placement, and also providing support to rural people working in large cities.

In Environment Development, there are initiatives for a clean and green village consisting of appropriate solid and liquid waste management, providing toilets in each household and in all public institutions and ensuring their proper use, tree plantation in accordance with local preference in homestead, schools and public institutions, watershed management especially renovation and revival of traditional water bodies, improving access to irrigation, Drainages, traditional drainages, rainwater harvesting, reducing local pollution of air, water and land, flood protection, etc. Besides these, there are preparations for disaster management in the form of subsidies for purchasing food as well as farm inputs, etc. for the helpless and disabled, and support for the renovation of the wells and renovation of the shops which were destroyed due to natural disasters.

The framework towards good governance is ensured through measures like strengthening of local democracy through strong and accountable Gram Panchayats and active gram sabhas, Provision of UIDAI cards to all, timely redressal of the grievances filed by the people such that: (a) Grievances of all nature to be submitted to the Gram Panchayats/Charge officers and dated receipts to be given, (b) Grievances to be redressed within three weeks along with written reply, (c) Institutionalization of regular open platforms for airing the grievances and their redressal, coordinated by the Gram Panchayats.

The promotion of healthy behavior emerges at a time when the world is ready to adopt a set of 17 goals 17 flagged as sustainable development goals (SDGs) rolled by the United Nations Development Programme, with the target year of 2030. SDGs were shaped by the churning of many conferences and summits, such as the Stockholm conference (1972) (UN 1972a) regarded as the first human environment conference, Earth Summit (1992), participated by 178 member countries adopted the agenda 21, which provides broadened action plan to cement partnership among the members to provide a way for sustainable development by conserving the environment (UN 1992a, b); Johannesburg declaration (2002), where global leaders reaffirm the commitment for human development by poverty reduction and environment protection with emphasis on multidimensional approach for partnership based on Millennium Declaration and Agenda 21 (UN 2002).

United Nations member countries adopted the resolution of the Rio + 20 Conference (2012), *"The future we want"*, which proposed a high-level political forum to leverage sustainable development based on the experiences of MDGs. The Rio + 20 conference gave the roadmap agenda, its implementation, financing, and support of small island developing states. To facilitate the outcome document of the Rio + 20 conference, the General assembly, United Nations 2013, set up an Open Working group of 30 member countries to produce a proposal document on sustainable development (UN 2013). The year 2015 became the landmark year for an international agency, as the SDGs were formalized, adopted, and ready to be implemented by the member countries in 2016 for the next 15 years.

In the Indian context, NITI Aayog has been entrusted with the role to coordinate work on Sustainable Development at the Central Level. The task at hand for the NITI Aayog is not merely to periodically collect data on SDGs but to act in a proactive manner so as to bring out the desired results not only quantitatively but also qualitatively. Ministry of Statistics and Programme Implementation (MoSPI) has undertaken a parallel to interact with the Central Ministries to evolve indicators that aptly reflect the SDG goals and targets.

At the State level, the Department of Planning and Development, Government of Bihar, is the nodal Department for the SDGs related work. Department of Planning and Development is also liaising with the NITI Aayog and other central bodies for taking forward the *Agenda 2030*. The Department of Planning and Development is entrusted with the task of preparing VISION and STRATEGY Document. It has completed the mapping of nodal departments, schemes goals/target wise. Presently, the Department of Planning and Development is in technical collaboration with Asian Development and Research Institute (ADRI) and United Nations Children's Fund (UNICEF) prepared a VISION and STRATEGY document for Bihar.

This book attempts to present a comprehensive evaluation of human development in Bihar through the lens of the Millennium Development Goals (MDGs). The diagnosis of the extent of problems is critical as the state embarks on the journey to the next level—The Sustainable Development Goals (SDGs). There have been shortcomings in a few areas and a retrospective evaluation becomes critical as the state of Bihar shoulders the responsibility to improve. This book emerges as an opportunity to draw upon the mistakes committed in the past and move ahead toward a better future with the Sustainable Development Goals (SDGs).

References

Sharma AN (2005) Agrarian relations and socio-economic Change in Bihar. Econ Polit Wkly 40(10):960–972. http://www.jstor.org/stable/4416306

Government of Bihar (2022). Ministry of Agriculture: About Us. Patna https://dbtagriculture.bihar.gov.in/krishimis/WebPortal/AboutUs.aspx#:~:text=Agriculture%20is%20the%20vital%20source,improved%20methods%20and%20system%20management

Mukherji A, Mukherji A (September 2012) Bihar: What Went Wrong? And What Changed? Working Paper No. 2012-107 New Delhi. National Institute of Public Finance and Policy. https://www.nipfp.org.in/media/medialibrary/2013/04/WP_2012_107.pdf: .

NITI Ayog (2008) Road map for rural industrialisation in Bihar. A report on the special task force on Bihar. Government of India, New Delhi. https://niti.gov.in/planningcommission.gov.in/docs/aboutus/taskforce/tsk_bri.pdf

Thapar R (1966) A history of India. Penguin Books, London

UN (1972a) Report of the United Nation conference on the human environment. United Nation, New York. Available on: https://undocs.org/en/A/CONF.48/14/Rev.1

UN (1992a) Report of the United Nation conference on environmemt and development. United Nation, New York. Available on: https://undocs.org/en/A/CONF.151/26/Rev.1

UN (1992b) United Nations conference on environment & development: agenda 21. United Nation, New York

UN (2002) Report of the world summit on sustainable development. United Nation, New York. Available on: https://undocs.org/pdf?symbol=en/A/Conf.199/20

UN (2013) Open working group of the general assembly on sustainable development goals. Agenda No. A/67/L.48/Rev.1. United Nation, New York

Chapter 2
Introspection of Human Development in India

AbstractBarmeier This chapter presents an appraisal of the programs and policies of the Government of India with regard to the achievement of the Millennium Development Goals. It lays down the challenges confronted in the process and provides recommendations to meet the gaps. Post-1990 witnessed the emergence of developing countries, which defined different programs to achieve human development under the Millennium Declaration in 2000. The literature review highlight suggests that the set of eight goals under the pack of millennium development goals, which are distinctly the most important agreement ever made for human development. MDGs succeeded in accelerating global political consensus for the common cause of human development by supporting ongoing projects, mobilizing aid, and improving dedicated targets. The implementation of SDGs is not an easy task and in India, challenges in the conceptualization and implementation of MDGs were faced at very distinct levels; goals, targets, and indicators.

Keywords India · MDG · Poverty · Health · Education · Environment

2.1 Introduction

At the dawn of the twenty-first-century global leaders united to address the global challenges of the current millennium. With the rapid expansion of the global population, there was a felt need to meet these challenges efficiently and effectively. These challenges manifest in the form of poverty, diseases, malnourishment, lack of education, gender inequality, environmental degradation, etc. In the developing and least developed countries, these challenges are even more prominent and have to be met on a priority basis. These challenges have been packaged in the form of Millennium Development Goals (MDGs) into a set of 8 goals, 21 targets, and 43 indicators with time-bound objectives. The MDGs act as a type of report card of the World's fight against global challenges. MDGs endeavor to generate incentives, particularly financial incentives, and develop global awareness, accountability, measurement metrics, and feedback mechanisms to meet global challenges.

U. K. Singh, *Human Development in Bihar, India*, Advances in Geographical and Environmental Sciences, https://doi.org/10.1007/978-981-19-3624-1_2

The origin of MDGs lies in the United Nations Millennium Declaration adopted by the General Assembly on 8th September 2000. "The declaration consists of 8 chapters and key objectives. It stresses values and principles such as peace, security and disarmament, development and poverty, eradication, human rights, protecting the environment, democracy, human rights, protection of vulnerable sections, ability to match the special needs of India, and strengthening the United Nations (UN 2000). The MDGs have specific targets and a timeline for obtaining them.

There are indications that a considerable amount of progress has been made in the achievement of MDGs, particularly by developing countries. The progress varies across regions, countries as well as goals. Some countries seem to be in a position to achieve all of the MDGs whereas others lag behind. It is expected that by 2015, most countries would have made substantial progress to achieve these goals. There can also be certain areas where considerable progress would not have been made by 2015 in the achievement of MDGs. This would represent a serious problem for some countries with a considerable number of poorer and vulnerable sections amongst them. Nevertheless, there seems to be a consensus amongst the policy-makers that significant progress would have been made by 2015 in the reduction of poverty, hunger, and disease.

India is a signatory to the millennium declaration adopted by the United Nations General Assembly in September 2000. India has shown its resolve to improve sanitation, health, education, dignity, financial inclusion, security, and reduce gender disparity. The Government of India (GoI) is also working to protect the environment with respect to air, water, soil, and the biosphere. The GoI is also focusing on growth with equity and sustainable development with a resolve to learn from domestic as well as international success stories.

It has been seen that the poorest states in India like Bihar, Orissa, Uttar Pradesh, Madhya Pradesh, etc. are among the most populous states of the country. These states also have the world MDGs indicators. Owing to rapid population growth, these states account for a larger share of the Indian population in 2015. Therefore, the performance of India in achieving MDGs goals would largely depend upon these states. Human development report 2020 shows that India has done well during the millennium development goals period as the share of the population living below the poverty line of USD 1.25 slipped to 32.7% in 2011 from 41.6% in 2005 (HDR 2020). Bihar is one of the poorest states in India with more than half of the people living below the poverty line. Bihar, Uttar Pradesh, Madhya Pradesh, and Rajasthan account for more than 50% of infant mortality in India.

It is an axiom that the problems facing humanity are closely entangled, and they tend to complicate the solution of one or more others. Conflicts and widespread diseases tend to thrive in regions, where people are poor and untrained, but they also act as commanding inhibitors of education and economic development. This shows the fundamental significance of a complete approach and a synchronized strategy. That was exactly the purpose of the United Nations Millennium Summit and Declaration. It becomes clear that a harmonized approach can yield much more than the sum of its parts. A synchronized strategy cannot be achieved without improved management among international institutions and agencies. This effort must mobilize the energies of all actors, including the private sector, philanthropic foundations, NGOs, academic and cultural institutions, etc. These realizations by the United Nations manifested in the form of Millennium Development Goals. To realize a given set of targets is not possible without the proper policy framework.

2.2 Literature Survey

Wilkinson and Hulme (2012) "state that while the MDGs spell out 'what is to be achieved' they do not explain 'why' so many people are poor what should be done and what policies ought to be persuaded to eradicate poverty. Any policy becomes pro-poor if the beneficiaries of the policies are poor". Ravallion and Chen (2003) connect the pro-poor policy to pro-poor growth and expressed it as the rate of change of poverty, the part of the distribution that is enhanced with improvement in some indicators of living standard. By this argument, we cannot expect a straightforward answer to human development. The answer lies in the explanation of the capability approach, which argued that any attempt to examine the poverty reduction of a particular cohort is done by evaluating the well-being (living standard, social welfare, or poverty level) of the said group. Sen (1979), connects human development with human capability and its functioning. He theorizes the capability approach as means for human development by merging two parts of capability valuable beings and doing (functioning). The capability approach becomes the core idea along with the discourse on the good life and basic conditions required for human development.

Human development became a major concern for policy-makers after the devastation of World War II when welfare became pivotal in policymaking across the globe. The concept of welfare and the role of government was advocated by John Maynard Keynes. Since the independence of India, the government has adopted pro-poor policies, starting from a planned economy to the affirmative provisions for marginalized and vulnerable cohorts, nationalization of banks to enhance access by farmers, and land reform to improve the productivity of the land.

All these steps are in line with the constitutional mandates where education, health, and welfare of the downtrodden is a state responsibility. To achieve harmony with the neoliberal regimes, India brought economic reforms in 1991, commonly known as LPG (Liberalization, Privatization, and Globalization) focused on the market economy. After a decade of the neoliberal era, at the dawn of the new millennium, world leaders tried to fill the gap between the developed and developing world through a set of goals in the form of MDGs (Lahiry 2014). The objective of MDGs is to improve human lives. It successfully changed the discourse among policy-makers from economic development to people-centric development, focused on poverty alleviation and hunger eradication (Fukuda-Parr 2004). MDG provides a new pattern of socioeconomic policies for human development, fundamentally it shows the common objective of equality, justice, and respect for human rights (Poku and Whitman 2011). Hulme and Wilkinson (2012) argued that MDGs spell out "What is to be achieved" but they do not explain "why" so many people are poor what should be done and which policies ought to be persuaded to achieve no poverty. McMichael et al. (2008) revealed that we cannot separate environmental causes from health causes, to eradicate poverty we must protect our environment, climate change accelerates the spread of infectious disease. The population growth creates pressure on natural surroundings and the rapid increase in global population and swift economic activities are potential threats to the biophysical and the ecological system across the globe. A study by Haines et al. (2007) established the adverse impact of energy sources on health issues. Limited access to affordable clean household energy translates into the risk of illness, morbidity, and mortality. Another study by Moyo et al. (2004) indicates that the microbiological quality of water declines between the collection source and utilization source in the household. The collected water is rarely treated for the purpose of dinking which attribute to water-borne diseases in the community, even households with piped water do not have the guarantee of safe water. Manning's (2009) scholarship critically examined the MDGs and shows the irregular progress of goals where some countries are experiencing robust growth while other countries are behind in the realization of many goals and targets.

India shelters nearly one-sixth of the global population with more than a billion population and therefore the progress made by India in the attainment of MDGs will considerably ease the burden of the global community in achieving the goals by the target year (Kaushal and Melkani 2005). At the national level, the progress of India is satisfactory but an examination at the state levels present a sorry picture of health and education, especially in the eastern and central India state namely Uttar Pradesh, Bihar, Jharkhand, Madhya Pradesh, Chhattisgarh, and Odisha (Hazarika 2012). In the past two decades, Bihar has tried to improve governance, strengthen the social infrastructure such as health and education, and worked for the betterment of the agriculture sector, which take the state on the path of robust development (Singh and Stern 2013). Kakwani (1980) has "derived a poverty measure similar to that of Sen by an alternative approach of transfer of income from rich to poor so that the income of every poor is brought to the poverty line".

This chapter analyzes the relevance and the role of the welfare social policy framework in shaping the trajectories of Bihar in coordination with central policies of India in order to achieve the Sustainable Development Goals. The following sections shall address the similarities and differences in policy-making to eradicate poverty and inequality at a cross-national level. Bihar has the potential to expand its economy and connect the marginalized section of the society with the mainstream economy and this book attempts to analyze the driving factors of welfare policy-making in all of them; how it ends up promoting a new model of development different from the west and according to their local requirements. At the time when India went for a poverty eradication programme, the major industrialized economies doubted whether inheriting such welfare policies would comply with the spirit of global market competitiveness. The benefits of central policies are realized in Bihar.

The factors range from the role of ideas, institutional factors, and the influence of political leaders which affect the development of social policy. Studies suggest that cohorts of ethnic minorities and marginalized remain in poverty for a long. Social constraints such as caste, religion, and ethnicity intensify the poverty risk and pass it to the next generation (Dang and Lanjouw 2015, 2018; Sumner 2013). In the context of India, Dang and Lanjouw (2018) argued that the escape from poverty is tougher for Schedule Caste and Schedule Tribe in comparison to other privileged castes. It aims to demonstrate how outcomes of social policies differ in countries of comparable economic circumstances. Social provisions in the global south became prevalent in the mid-1990s in the form of anti-poverty cash transfers or subsidies which were financed by taxes and were non-contributory in nature. The development of the Minimum Living Standard Guarantee programme or Dibao in urban China in the 90s or cash transfers based on school attendance (Bolsa Escola) in Brazilian cities in the 90s or child support grants (1998) in South Africa are prominent examples (Barrientos 2013). Social assistance through cash transfer initiatives of diverse countries from Pakistan to Argentina, Kenya, Nigeria, etc. (Barrientos et al. 2010). Many countries also brought in policy initiatives which focused on capability building of the poor along with income transfers such as India's MNREGA while some brought in social pension or health insurance benefits to a wider population. The economic crisis of the 80s and 90s, which brought in economic reforms of LPGs initially caused income inequality and poverty to rise. Barientos and Hulme (2009) put it this way: "The poverty and vulnerability which arose from globalization and economic reforms are key drivers for social protection". However, the resulting growth out of those structural transformations in the emerging economies led to an increase in the availability of resources for social schemes. Even international organizations such as the World Bank and domestic governments started supporting new forms of social protection. There was this popular discourse of inclusive growth doing the rounds on global platforms, which caused domestic politics to take a back seat from the explanations of the origins of social policy initiatives.

Mkandawire (2004) and Adesina (2009) however argue that the mainstream discourses on development in the global north used social policy only as a corrective measure for economic fiascos rather than as a concentrated form of public action to improve standards of living. Midgley (2004) argues that such a practice didn't take into account the collectivist nonwestern welfare traditions based on culture and local practices of the domestic domain. This is why the West failed to understand the structural challenges faced by the global south because they were viewed from the lens of those faced by the western advanced countries. Rodrik (2008) says that the North's main concern was that "international economic integration does not lead to domestic social disintegration". This was being touted as a possible reason for governments in the middle and lower-income countries to enlarge welfare provisions. Subramanian (2013) similarly observed with respect to India that financial integration ultimately puts the economy at greater risk and therefore leads to an increase in fiscal spending and greater deficits.

Therefore, Barrientis and Hulme suggest that a more local based instead of a globalized policy framework has been at work. Historical trajectories and other political aspects at home are significant in shaping social policy interventions with respect to bigger economic changes. Also, in the global south like China, India, Brazil, and South Africa; policies have a dual function—both productive as well as protective. Such policies turn out to be game-changers in the long-run economic growth as well as have a mass political appeal for leaders of all kinds of regimes. Dréze and Sen (2013) said how policies in the global south have an element of a social investment regime, i.e., it's an election promise or that shall give the party or a leader returns in terms of electoral power. Harris (2013) and Jayal (2013) mention how India's focus has also been on strengthening social rights. The final inference which may be drawn here is that social policies and their extent is highly contested and subject to debate in most of these nations. This also leads to the fact that domestic players which include individual leaders, political parties, civil society, institutions, and judges have an influential role to play.

This paper thus takes a comparative analysis approach to figure out how the welfare policies are shaped in these countries of the global south taking into account the domestic factors as causes for this programme. Initially, these countries were driven by the legacies of global liberal economics of the 90 s that had brought down social sector investment but new social policies were encouraged by the resultant faster economic growth.

2.3 Millennium Development Goals (MDGs)

MDGs are set of eight goals dedicated to human development by leveraging the social goals, ranging from poverty reduction to dealing with mortality incidences and reversing the spread of communicable diseases, maintaining the balance with nature, and mobilizing partnership for development to be achieved by the target year 2015. The development framework in a set of MDGs is agreed upon by the member countries, international lending agencies, and other international and regional organization. Together they have mobilized resources to prepare the world to walk on the path of human development.

United Nations along with institutions like the World Bank Group (WBG) and International Monetary Fund (IMF) are pledged to provide funds in order to meet the financial requirement of member countries. In the passage of time UN and WBG together extend their coverage, and provide umbrella coverage to policies related to sustainable development, in line with human development. From the experience of MDGs, UN-WBG bonding stretches its coverage for financing the post-2015 agenda, together they are financing agriculture projects, climate change, and resilience programmes, covering disaster risk management, and promoting the social programmes related to food security, gender parity, health infrastructure, education, affordable and sustainable energy source and commerce activity (IMF 2021).

The WBG has distinct institutions for the smooth flow of funds and dispute settlement. The World Bank's synonym International Bank of Reconstruction and Development (IBRD) along with the International Development Association (IDA) work for financing social programmes in the lower- and middle-income group of countries. International Finance Corporation (IFC) deals exclusively with the private sector and its services include investment financing; capital mobilization for the international financial market; advisory services for government and private entities. Multilateral

Investment Guarantee Agency (MIGA) works to improve the lives of marginalized and vulnerable populations in the developing world by facilitating funds for economic growth and poverty reduction. In case of any investment dispute, the International Centre for Settlement of Investment Dispute (ICSID) provides arbitration and conciliation service.

2.4 Goals and Performance

2.4.1 Goal 1: Eradicate Extreme Poverty and Hunger

In the year 1990, almost 50% of the population in the developing regions lived on fewer than $1.25 a day and the largest part of people living on lower than $1.25 a day belonged to two regions: Southern Asia and Sub -Saharan Africa. In the year 2010, one-third of the world's 1.2 billion extreme poor resided only in India. China, even after its progress in poverty reduction ranked second with 13% of the global extreme poor followed by Nigeria (9%), Bangladesh (5%), and the Democratic Republic of Congo (5%). Close to two-thirds of the extremely poor reside in these five countries for the year 2010. Table 2.1 depicts the targets and indicators of MDG1.

Table 2.2 depicts the absolute number of people living in extreme poverty decline in India. The incidence of poverty fell from 59.8 to 51.3% between 1981 and 2005, again in 2005 in India it comes to 41.6 which is far behind the MDGs target of 25.7

Table 2.1 MDG1 (targets and indicators)

Reduce the share of the poor population (income <= $1) by half by 1900–2015	Share of population living with income less than <= $1.25/day
	Poverty intensity
	Proportion of marginalized section in national consumption
Attain holistic employment and a handsome job for all	Growth rate of GDP/ employed person
	Proportion of employment and population
	Share of employed population living below income <= $1/day
	Share of the population having their own account and their contribution towards their family
Reduce the share of population affected by hunger population by half by 1900–2015	Incidence of under-weight children (under five years)
	Share of population below dietary energy consumption

Source UN 2000

Table 2.2 Population living below $1.25/day (India and Bihar) and rate of change to meet the target

Region	Population living below $1.25/day					Target	Annual rate of change	Required rate of change	Required annual rate of change
India	1981	1990	2005	2011	2012	2015	1990–2005	2005–2015	2005–2015
	59.8	47.8	41.6	–	23.6	25.7	−1.4	−15.9	−4.8

Source World Bank (1981–2015)

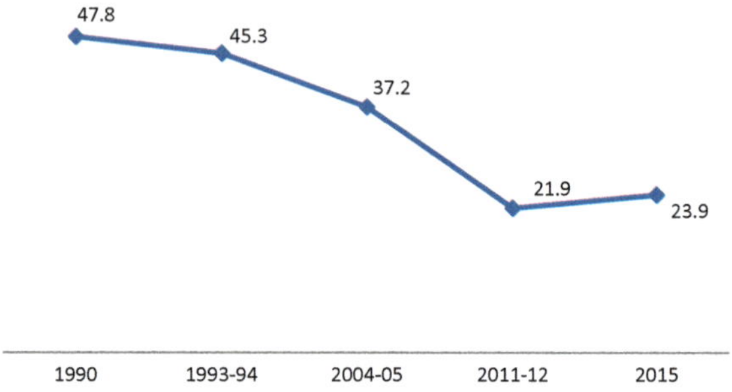

Fig. 2.1 Poverty head count ratio, India. *Source* Planning Commission of India (2015)

for India. On this basis, it can be estimated that India needs the −4 annual rate of change to meet the target.

Figure 2.1 shows that India's poverty headcount ratio is declining with rapid speed, i.e., the annual rate of percent change is 1.9% between 2004 and 2012 compared to 0.7% from 1993 to 2004. The country has achieved the target in 2012 with a 23.9% headcount ratio compared to 47.8% in 47.8 (Planning Commission of India 2015).

2.4.2 *MDG2 Achieve Universal Primary Education*

Goal 2 as reflected in Table 2.3, aims to achieve a full course of primary schooling for all girls and boys by 2015. "As of 2001 estimates around 115 million children of primary school age and the majority of them, girls do not attend school" Wright and Millward (2005) .

Table 2.3 MDG2: targets and indicators

Ensure attainment of primary education coverage by 2015	Net primary enrolment
	Share of beginners who completed the last grade of primary
	Literacy rate of adults (15–24) year of age

Source UN 2000

Table 2.4 Net primary enrolment (India)

Region	1990	Target (2015)	2011
India	77	100	100

Source MDG Country Report (2014)

Education is one of the primary and foremost stepping stones in nation-building. India has dedicated programs for imparting free and compulsory education to all children. With this aim, the Indian Parliament has enacted legislation making free and compulsory education a right of every child in the age group 6–14 years, which came into force on 1st April 2010. The main aim of the 'Right to Education' enacted by the Government is to ensure universal education for all children aged 6–14 years by making it free and compulsory for this age group (RTE Act 2009). Envisioned within this right to education is the hope that education would promote the values of human dignity, equality, human rights and freedom, non-racism, and non-sexist and the Ministry of Human Resource and Development in India is responsible for implementing the Millennium Development Goal 2.

The net enrolment ratio (NER) in primary education is the most important indicator to evaluate whether the country is likely to achieve the 2015 target of universal primary education for all children aged 6–10 years. As shown in Table 2.4, India has achieved a target of 100 in 2011.

2.4.3 MDG3 Gender Parity and Women Empowerment

In order to achieve the target of gender parity and women empowerment, education is the only way. Gender gaps in attaining education have reduced, but disparities remain among regions in all levels of education, particularly for the most excluded and marginalized. Gender parity in primary schooling worldwide has been achieved. There has been major growth across all developing and least developing countries in narrowing gender gaps in primary school attendance. However, girls continue to face restrictions on schooling, particularly in Northern Africa, Sub-Saharan Africa, and Western Asia. Although more girls are now in school in Sub-Saharan Africa, only 93 girls are enrolled in primary school for every 100 boys (MDG 2000) (Table 2.5).

Table 2.5 MDG3 targets and indicators

Achieve gender parity at primary and secondary levels by 2005 and in all stages of education by 2015	Gender ratio in primary, secondary, and tertiary level of education
	Proportion of women in paid employment in the nonfarm sector
	Proportion of women national parliament

Source Millennium Declaration (2000)

The next important indicator for Target 4 under MDG3 is the women's share in Wage Employment in the Non-Agricultural Sector, which is given as the proportion of female workers in the nonagricultural sector defined as a percentage of total employment in the sector. This computes the extent to which labor markets are welcoming to women in the industry and service sector, which affects not only equal employment opportunities for women but also the economic efficiency bestowed upon by the flexibility in the labor market and reflects the economic factors in social empowerment of women. It is a substance of delay in time to get the full effect of gender equity in education on women's participation in the labor markets of industry and services in India. The rate of change over time in terms of the share of women in wage employment in the non-agricultural sector is passive. The Periodic Labor Force Survey (PLFS) (2019) reported that the percentage of females in the wage employment in the nonfarm sector was alarming. During the survey period 2017–18, the percentage of female participation in the labour force in urban areas declined by 1.9% from 17.8% in 2004–05 to 15.9% in 2017–18, during the same period rural area women's participation reduced by 15.1% from 33.3 to 18.2%. Figure 2.2 depicts the story of female share in wage employment in the nonagriculture sector.

The third indicator of target 4 talks about the proportion of women in the national parliament. As per the data released by the Inter-Parliamentary Union (IPU), India occupies the 149th position in the world for the proportion of women in decision-making in the National Parliament. Table 2.6 depicts the increasing percentage of women in the Indian Parliament but not with satisfactory speed. Data shows that in the Lower house, the representation of women increased to 14% in the 17th Lok Sabha held in 2019 from 8.3% in the 15th parliamentary election concluded in 2009 (Economic Times, 2019), while the numbers of women in the upper house remain stagnant (IPU 2015).

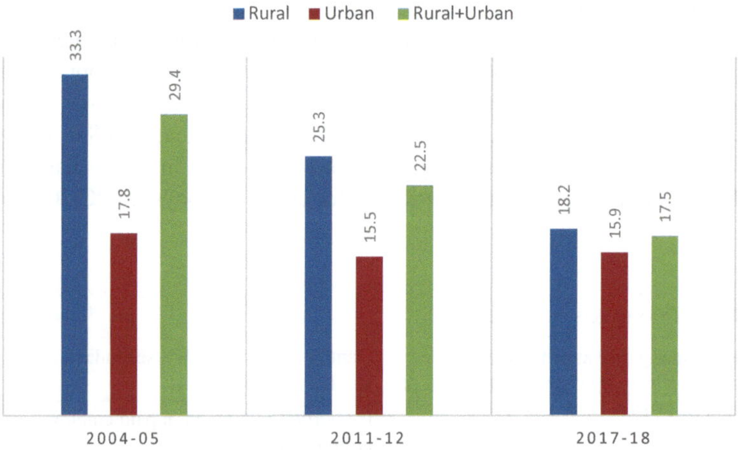

Fig. 2.2 Proportion of Female in PLFS, India. *Source* International Parliamentary Union

Table 2.6 Proportion of women in India's National Parliament

Country	Lower house				Upper house				Global ranking
India	Election (Lok Sabha)	Seats	Women	Women's share (%)	Date of election	Seats	Women	Women's share (%)	
	15th	545	52	8.3	06/2004	250	28	11.2	104
	16th	542	64	11.8	01/2014	244	27	11.1	149
	17th	542	78	14	01/2018	244	28	11.5	149

Sources Inter-Parliamentary Union (2015)

2.4.4 Goal 4: Reduce Child Mortality

The level of child mortality is a reflection of a country's health status in general and, specifically, of the health status of children. It is also a reflection of the quality and efficiency of the health system operating in a particular country (Mckerrow and Mulaudzi 2010). Case et al. (2002) revealed the casualties between the level of income and status of children's health, depending on circumstances in developed and developing countries, the relation shows the multidimensional relationship between health and income.

Bryce et al. (2003), argued that success in MDG4 (child mortality) can be achieved by enhancing access to sustainable health facilities to a large proportion as the survival of a child is dependent on health care received by mother and child. Child mortality reflects the advancement of a particular country in medical science and its accessibility to various groups of society. Further, studies have explored the economic disparity among the subgroups within the country and suggest that governments lacking resources would not be able to halt the mortality rate as a focused health intervention, foreign aid, and advanced technology are required to curb child mortality incidence.

MDG4 has recorded outstanding progress in achieving its target; it still falls short of what is needed to attain the global target of 29 deaths per 1000 live births. The global under-5 mortality rate (U5MR) slides from 87 deaths per 1000 live births in the year 1990 to only 51 in the year 2011 amounting to a 41% decrease. In 2011, U5MR in developing regions has 57 deaths per 1000 live births which was eightfold the rate in developed countries (about 7 per 1000). Kumar and Singhal (2020) reported that the Under-5 Mortality Rate in India reduced to 42.4 deaths/1000 livebirths in 2017 from 83.1 deaths/1000 in 2000. The neonatal mortality declined to 23.5 deaths from 38/1000 per live birth (Table 2.7).

Although the countries of North Asia, Latin America, East Africa, Caribbean, West Asia, and Southeast Asia have successfully reduced the under-5 mortality incidences, the sub-Saharan countries have experienced a massive decline in mortality cases, and the region has doubled the annual rate in reduction between 1990–2000 and 2000–2011. UNICEF (2020) reported that an estimated 53% (2.8 million) of under-5 death incidences happened in the Sub-Saharan region and nearly 1.5 million died before attaining the age of 5 in central and South Asia. Collectively both regions accounted for 80% of cases of under-5 mortality. India, Nigeria, the Democratic Republic of Congo, and Pakistan are collectively responsible for the world's half under-5 deaths. India performed worst with 24% succeeded by Nigeria with 11%

Table 2.7 MDG4 reduce child mortality

Under-5 mortality rate should be decreased by two-thirds between 1990 and 2015	Under-5 mortality rate
	Infant mortality rate
	Ratio of 1% of children immunized for measles

Source UN 2000

sharing one-third of total under-5 child mortality cases. UNICEF (2012) reported that slower mortality of neonatal as compared to the mortality rate of older children translated into a decline in mortality cases, registering a 17% rise since 1990.

Awareness and removal of social restrictions through educating community leaders and family members are key to achieving MDH4 (Lahariya and Paul 2010). The study by Odus et al. (2021) established the relationship between deaths and care provided by the antenatal care centers, 99% of death related to women who visited antenatal centers less than 8 times while one percent lost their life who visited more than 8 times. Though, in India, the share of women who had an antenatal visits in the past days of pregnancy increased to 51% in 2006 from 44% in 1992, in the same period delivery of medical assistance improved to 49% from 35% (NFHS3 2007). IIPA (2017) report revealed that the share of women who had received antenatal care rose to 84% in 2015–2016 from 77% in 2005–2006, and antenatal visits in the last days of pregnancy increased to 51% in 2006 from 44% in 1992. There is an urgent need to strengthen health infrastructure in order to attain the goals.

The Third Indicator supplies a measure of the coverage and the quality of the child-health care system in the country. The only indicator for immunization of children that were selected in MDGs was the Proportion of 1-year-old children immunized against measles the discernment that its level of coverage is likely to represent coverage by other antigens like BCG, DPT, and polio as well, as they are provided prior the antigen of measles could be given. Besides, within the category of vaccine-preventable diseases of childhood, the leading cause of child mortality remains—Measles. The aim is to achieve immunization against measles above 90% (MDG 2000). Although, evidence suggests that the countries need to steadily move forward to achieve the target of measles immunization estimated 62% of children age up to 12 years received a basic vaccine (IIPA 2017). Figure 2.3 reflects that 81.1% of children were immunized against measles during NFHS 4 (2015–16) while in the same period in the urban areas the percentage is 83.2% and 80.3 in rural areas. There is a significant rise in coverage from the period of NFHS3 (2005–2006).

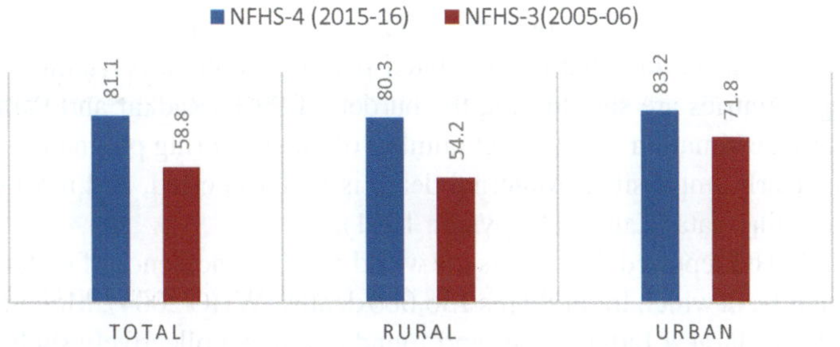

Fig. 2.3 Immunization against measles in India. *Source* NFHS4 and NFHS3

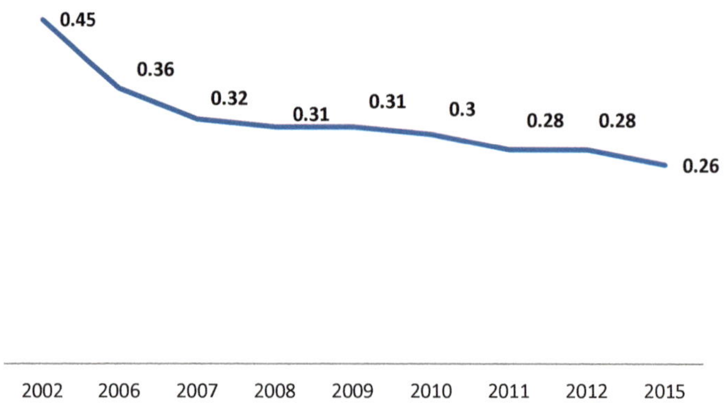

Fig. 2.4 Trends of HIV incidence in India. *Source* MOSPI (2017)

2.4.5 *Goal 5: Improve Maternal Health*

Lack of medical facilities is the main reason behind maternal mortality. Women die from the problem which can be prevented by the use of medicine and medical help. Health complications before pregnancy become severe and life-threatening during pregnancy. WHO (2019) identified four major problems that accounted for 75% of MMR, which are

i. Excess bleeding,
ii. Infections,
iii. Increased blood pressure,
iv. Delivery complications, and
v. Unsafe abortion.

The 1994 Cairo declaration of the international conference on Population and Development (ICPD) envisaged women's reproductive health. The issue of women's health is a major concern for human development. No country has achieved success in fulfilling the targets set in the conference. The situation of maternal health is worse in the developing world. Pregnancy-related death accounted for almost 600,000, developing countries are shouldering the burden of 99% (Radkar and Parasuraman 2007). Annual estimation of the exact number of deaths during pregnancy and child delivery is nearly impossible. Maternal death is underreported, and most countries are not providing data (Zahr and Royston 1991).

In 2004, WHO reported that across the world 536,000 incidence of maternal death occurs annually, in which India shares 136,000 deaths (WHO 2004). Mohanty (2012) surveyed the 18 largest Indian states and found that they collectively sheltered 95% of the Indian population. This study cemented the argument of uneven maternal care facilities and that access to antenatal care varies from one state to another. Bihar lies at the bottom with 17% while Tamil Nadu was the top performer with 96%. MMR can be controlled with effective healthcare solutions. There is a need to improve the access to better health services for women during pregnancy and after delivery of the

Table 2.8 MDG5, targets and indicators

Decrease the maternal mortality by three quarters between the 1990 and 2015	Ratio of maternal mortality
	Proportion of births attained by a trained health worker
Attain the universal health access by 2015	Contraceptive use rate
	Adolescent birth rate
	Coverage of antenatal
	Unmet call for family planning

Source UN 2000

child. Maternal health is directly proportional to child health. To improve maternal health, there is a need for poverty reduction, expansion of health infrastructure, and awareness campaigns.

The second target of MDG5 is the expansion of women's reproductive health services by 2015. Reproductive health deals with the right of women to their own bodies. It is not merely about the absence of diseases or infirmity but encompasses mental, physical, and social behavior which influences the sexual life of women. Use of contraceptives such as condoms and pills, gap between two deliveries, etc. are parameters. The final country report of India reflects that the use of contraceptives in India at the national level is 5.5% (2015), which improved slightly from 2005 to 06 (MOSPI 2017) (Table 2.8).

The first indicator of Goal 5 is the maternal mortality rate which is defined as the number of women who die from any cause that is related to or aggravated by the pregnancy or its management (not including accidental or incidental causes) during the pregnancy and childbirth or after only 42 days of the termination of the pregnancy, notwithstanding the duration and site of pregnancy, per 100,000 live births. WHO (2004) reported that the burden of maternal mortality is carried by 13 countries and the World Health Organization (WHO) collectively accounted for 67% of total maternal deaths. With 136,000 incidences, India leads the list followed by 37,000 (Nigeria), 26,000 (Pakistan), 20,000 (DR Congo), 24,000 (Ethiopia), 21,000 (Tanzania), 20,000 (Afghanistan), 16,000 (Bangladesh), China, Angola and Kenya (11,000 each), and Kenya (11,000).

Significant improvement was registered in the global maternal mortality rate (MMR) during the period of MDG with decreased death incidence from 330/10,000 to 210/10,000 (UN 2017). From an estimated MMR level of 540 per 100,000 live births in 2000, India is required to reduce the MMR to 109 live births by 2015. At the historical pace of decrease, India lags by 31 points (MOSPI 2014) to achieve 38 lives per 100,000 live births. Though, SRS (2016–18) reported that maternal death in India decreased to 113/100,000 in 2016–18 from 130/100,000 in 2014–16 (PIB 2021).

The second indicator shows births attended by skilled health personnel, given as the proportion of total live births that are attended by a skilled birth assistant trained in providing lifesaving obstetric care. Safe motherhood depends mainly on delivery by trained/professional personnel, particularly through institutional facilities. Among other things, ensuring ante-natal care of prospective mothers at health centers and recommended doses of IFT are important factors that help improve maternal health and reduce life risk during pregnancy.

The National Family Health Survey (2015–16) revealed that institution delivery in India increased to 79% in 2015–16 from 39% in 2005–2006 (IIPA 2017) As per the Coverage Evaluation Survey (CES), 2009, the status of delivery attended by skilled personnel is 76.2% which was 47.6% as per District Level Household Survey (DLHS 2002–04). With the existing rate of increase in deliveries by skilled personnel, the likely achievement, is far short of the targeted universal coverage. The second target of MDG5 is the expansion of women's reproductive health services by 2015. Reproductive health deals with the right of women to their bodies. It is not merely the absence of diseases or infirmities but encompasses mental, physical, and social behavior which influences the sexual life of women. Use of contraceptives such as condoms and pills, gap between two deliveries, etc. are parameters. The final country report of India reflects that use of contraceptive in India at the national level is 5.5 (2015) percent, which improved slightly from 2005 to 06 (MOSPI 2017).

2.4.6 Goal 6: Reverse the Spread of Communicable Diseases (HIV/AIDS, Malaria and Other Diseases)

Kadiyala and Barnett (2004) revealed that tuberculosis (TB) and malaria are responsible for maximum deaths, while ignorance makes AIDS the leading cause of death in India. AIDS cases are directly proportional to poverty and inequality (Piot et al. 2007). HIV data are bone-chilling. 37.7 million of the world's population live with HIV, out of this 6.1 people are not aware of their HIV infection, and only 84% of people have correct knowledge about their HIV status (UNAIDS 2021).

During the period of MDG, new infections were controlled and the number of infected cases was reduced to 2.1 million in 2013 from 3.5 million in 2000. Coverage of anti-retroviral therapy increased to 13.6 million (2013) from 80,000 (2003), halting the death incidence by escaping 7.6 million deaths from 1995 to 2013 (UN 2016).

After HIV, malaria posed a great threat to human development. Across the world, 85% of communicable diseases are burdened by malaria along with HIV, diarrhea, TB, measles, hepatitis B, and pneumonia. South east is the global hotspot of malaria which contribute to 2.5 million cases, with a 76% share in the region's contribution (Kumar et al. 2007) (Table 2.9).

The first target is to halt and reverse the spread of HIV/AIDS. Estimation of the total number of People Living with HIV (PLHIV) is a useful indicator for assessing the severity of the epidemic at a particular point in time or its trend over the duration of time. As per the UN report, India has the third highest number of people living with HIV in the world with 2.1 million Indians accounting for about four out of 10

Table 2.9 MDG6 targets and indicators

Stopped the spread of HIV by 2015 and started the reversal of HIV/AIDS	Incidence of HIV under age group 14–24 year
	Use of condom during last high-risk sex
	Number of people having correct knowledge of HIV/AIDS
Ensure Universal availability of health facility for HIV/AIDS patients	Orphans and no-orphans proportion of school attendance aged 15–24 year
	Proportion of people having advanced HIV infection and access to antiretroviral drugs
	Number of deaths due to malaria
Stopped the spread of HIV by 2015 and started the reversal of malaria and other diseases	Proportion of under-five year of children sleeping under bed nets
	Proportion of under-five year of children covered by anti-malaria drug
	Number of deaths due to tuberculosis
	Proportion of confirmed tuberculosis cases and cured under short course medication

Source UN 2000

Table 2.10 Status of HIV/AIDS in India

	Prevalence of HIV among adults			Use of condom during last high-risk sex		Number of people (15–24 years) having correct knowledge of HIV/AIDS		
India	2007	2012	2015	2010	2010	2015		
						Women	Men	Youth
	0.34	0.28	0.26	54%	74%	21.5	31	26.5

Source Country report 2017

Table 2.11 Incidence of tuberculosis in India

	Incidence (per lakh)	Prevalence (per lakh)	Mortality (per lakh)
1990	216	465	38
1995	216	465	38
2000	216	438	39
2005	209	365	36
2009	190	289	29
2010	185	269	27
2011	181	249	24
2014	167	195	17

Source Ministry of Health and Family Welfare

people infected with the deadly virus in the Asia–Pacific region (The Hindu, 17 July 2014).

As depicted in Table 2.10, trends in HIV prevalence among persons aged 15–24 years could be used as a reasonable proxy indicator for the new infections in the population.

In India HIV/AIDS continues to decline at the national level with an overall reduction in adult HIV prevalence, HIV incidence (new infections), and AIDS-related mortality in the country. The latest HIV estimates provide sound evidence of the current trend of the disease. There is a massive reduction of 66% of cases of new HIV infection in 2015. An estimated 75,948 cases were registered in 2015 against 106,335 in 2007. The trend of annual AIDS deaths is also showing a steady decline since the rollout of the free antiretroviral treatment (ART) program in India in 2004. However, the HIV prevalence for adults at the national level has steadily declined from an estimated level of 0.34% in 2007 through 0.28% in 2012 to 0.26% in 2015. Similar regular declines are noted among both men and women at the national level (Country Report 2017).

Figure 2.5 demonstrates the prevalence of HIV among the young Indian population (15–24 years) at the national level which has declined from 0.15 to 0.11% (from the year 2007 to 2011) (Department of Aids Control India 2012). Unlike adults (15–49 years), HIV prevalence where HIV prevalence among males is around 1.5 times as compared of females, in the young (15–24 years) population, HIV prevalence is equal among men and women at 0.11% in 2011. On condom use at high-risk sex, the Behavioral Surveillance Survey (BSS) was organized to monitor the changes in different risk groups with respect to HIV/AIDS for knowledge and behavior parameters, indicating the prevalence of condom use among nonregular sex partners. According to BSS conducted in 2001 2006, the national estimates for condom use at last high-risk sex (percent)—the proportion of the population aged 15–24 years who used a condom during last sex with a non-regular partner witnessed

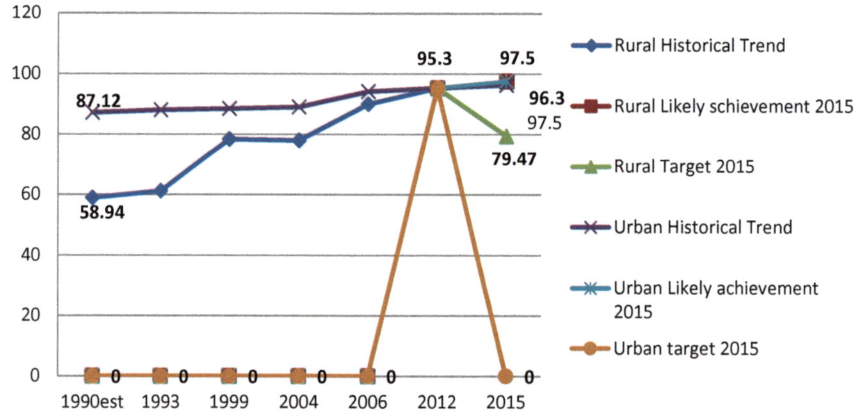

Fig. 2.5 Household's having access to improved sources of drinking water. *Source* NSSO (2013)

a 19% increase from 51.9% in 2001 to 61.7% in 2006. Sex education is a hot topic of discussion that whether to give sex education in schools or not on the proportion of the population with comprehensive correct knowledge of HIV/AIDS Government of India (2018) report exposed the proportion of sex education among the 15–24 years group, the national estimate for the proportion of the population aged 15–24 years with comprehensive correct Knowledge of HIV/AIDS is 26. 5% in 2015–16.

The second target of Goal 6 is universal access to HIV/AIDS treatment for all those who need it and its indicator for the proportion of the population with access to antiretroviral drugs therapy who have advanced HIV infection; it indicates the percentage of all people living with HIV who are receiving antiretroviral. The number of people getting this therapy is continuously increasing in India. It rose from 26% in 2011 to 36% in 2013 in India and they are likely to achieve the target of 2015 (World Bank 2014).

The third target is to overturn the incidence of malaria and other major diseases like tuberculosis (TB) and halt it by 2015. Globally 3.2 billion people are at high risk of contracting malaria. An estimated 198 million cases were detected and killed approx. 584,000 people, most of them are children under-five age from Africa in 2013 (WHO 2014). The number of deaths per 100,000 caused by malaria in India halved between 1990 and 2013 but still, it remains a remarkable health burden for Indians. About 61 million cases of malaria were detected in 2013 which took more than 116,000 lives (The Hindu 2014).

Tuberculosis is a major epidemic in developing and least developing nations with an estimated 9 million cases of TB reported in 2013 (including 1.1 million cases among people with HIV) and 1.6 million deaths (including 360,000 people with HIV), proving this epidemic as world's biggest infectious murderer (WHO 2014).

Assessment of the WHO report (2015) reflects the tuberculosis (TB) incidence in India. India had the largest number of TB cases in the world, 23% of global cases.

Table 2.11 explains the prevalence rate of TB in India which has come down from 465 per 100,000 lives in 1990 to 249 per 100,000 lives in 2011 to 167 lives in 2014 and further Table 21 gives the trends of the tuberculosis situation in India, the death due to TB has reduced to 17 per 100,000 lives in 2014 from 38 per 100,000 lives in 1990 (WHO 2015).

2.4.7 Goal 7: Ensure Environment Sustainability

MDG7 concentrates on ensuring environmental sustainability, and its targets and indicators are reviewed within the framework of other local, national and global development initiatives. After the publication of the Brundtland Commission Report 1987, the idea of sustainability across various sectors has gained popularity; the formulation of the 2000 United Nations MDGs (particularly MDG1 and MDG7, which articulate poverty eradication and environmental protection respectively) and the 2012 Rio + 20 conference on sustainable development (Rio + 20, June 2012). Environmental sustainability is a prerequisite for poverty alleviation and sustainable development. The developing countries' major challenge is to achieve environmental sustainability and human development in spite of macroeconomic development. India's major development has been measured by addressing poverty and inequality by committed service delivery programs targeting the poor such as drinking water, sanitation, electricity, and solid waste management (Table 2.12).

The first target of MDG7 is to provide a framework of policies and programs integrated into the principles of Sustainable Development and reverse the loss of environmental resources. Forests are a safety net for the poor, but they continue to disappear at an alarming rate. Conservation, expansion, and improving the quality of forests have enormous domestic and global benefits. India has identified it and takes it on priority in national policy formation.

India shows some progress toward the Seventh Millennium Development Goal of ensuring environmental sustainability. Table 2.13 characterize the total forest cover of India in very dense, moderately dense, and open forest, as per the MDG country report 2017, there is an increase in the total forest cover of India from 2013 to 2015

Table 2.12 MDG7 targets and indicators

Holistic policy for environment conservation	Ratio of land and forest cover
	Per capita and per GDP CO_2 emission
	Intake of ozone-depleting substance
Preserve biodiversity by a considerable reduction in the rate of loss by 2010	Fish stock under safe biological limit
	Proportion of consumed water resource
	Proportion of protected marine and terrestrial area
	Number of endangered species
Improve the accessibility of safe water for drinking purposes by 2015	Better water facilities for a larger proportion of population
Ensure better life for 100 million slum dwellers by 2020	Sanitation facilities for a larger proportion of the population
	Slum-dwellers in urban areas

Source UN 2000

Table 2.13 India's forest cover

Class	Area (km^2)	% age of geographical area
Very dense (density above 70%)	85904	2.61
Moderately dense (density 40 <= 70)	315374	9.59
Open forest (density 10 <= 40)	300395	9.14
Total	701673	21.34

Source The India State of Forest Report (ISFR) 2015

by 3775 km^2. In 2015, the total forest cover of India is 701673 km^2, which is 21.34% of the total geographic area of India.

The second indicator of goal 7 is CO_2 emission, which stems from fossils and is inclusive of CO_2 emission from the consumption of solid, liquid, and gas fuels and gas flaring.

Ozone-Depleting Substances (ODS): It contains chlorine or bromine, most of the biologically damaging ultraviolet radiations, which deteriorates the stratospheric ozone layer that absorbs them. India has committed to phasing out ODS in a phased manner since 1990 with the production and consumption of CFC to be completely phased out by 2010, methyl bromide by 2015, and HCFCs by 2040 (GoI 2015).

Target B is about the significant rate of reduction in biodiversity loss by 2010. Biodiversity is vitally important; billions of people rely directly on diverse species for their livelihoods and survival. As per the United Nations report, the member countries missed the 2010 target to slow the pace of reduction in biodiversity. At present, an estimated 17,000 species of plants and animals are at risk of extinction, and this number of in-danger species is increasing day by day. The main causes of biodiversity loss are due to high rates of consumption, habitat loss due to invasive species, pollution, and climate change that are not being sufficiently addressed even with increased investment (MOSPI 2010).

Target C of Goal 7 aims to reduce the proportion of people without sustainable safe drinking water by half by 2015. The member countries achieved this target 5 years ahead of schedule. An estimated 2.3 billion people globally are now receiving water through improved drinking water sources but even in 2012, 748 million people were still deprived of an improved drinking water source. One-fourth of the world's population got better sanitation since 1990. But the problem of open defecation still prevails and around one billion people still defecate open and 82% are from middle-income populous countries and around 2.5 billion people still do not get improved sanitation facilities. Every target of the MDGs rides on the achievement of water and sanitation targets such as eradicating extreme poverty and hunger, achieving universal primary education, promoting gender equality and empowering women, reducing child mortality, improving maternal health, combating HIV, AIDS, malaria, and other diseases, and ensuring environmental sustainability.

India has performed well on the access of improved methods of drinking water, which include bottled water, piped water, and dwelling piped water to yard/plot, public/tap/standpipe, tube well/borewell, protected well, protected spring, and rain-water collection. An estimated 88.5% rural population and 95% of urban households had improved sources of drinking water. The current trends suggest (Fig. 2.5) that India will achieve its target by 2015 in both rural and urban areas and it is not wrong to say that India already achieved its target (Census 2011).

The sanitation facility is directly proportional to living conditions and strictly attached to the health and hygiene of the population. WHO, through Global Water Supply and Sanitation Assessment 2000, interpreted sanitation as including a connection to a septic tank or sewer system, pour-flush latrine, or simple-pit or improved ventilated pit latrine with allocation for improved local technologies (WHO 2000).

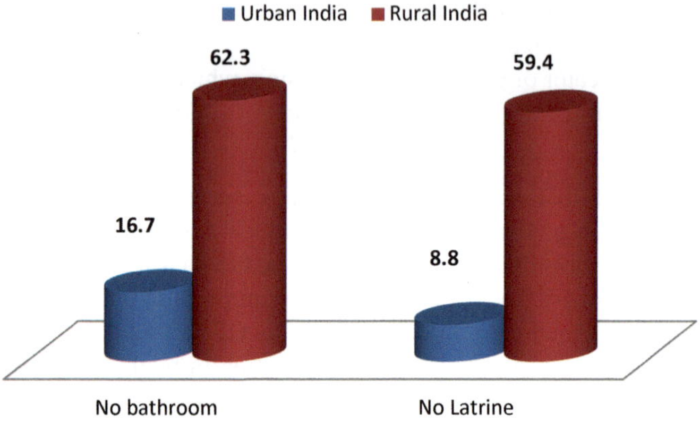

Fig. 2.6 Household's having bathroom and latrine, India. *Source* NSSO (2014)

31.9% of rural and 85.3% of the urban population of India in 2008–2009 had access to improved sanitation which became 38.8% in rural and 89.6% in urban areas and is likely to achieve the target of 100% in 2015. Figure 2.6 depicts that in rural areas only 37.7% and in urban areas 40.6% of people had access to bathrooms and the condition for use of latrine is not also good data shows that 83.3% in rural and 91.2% in urban areas had access to latrine during 2012. It shows a wide gap between rural and urban India in respect of bathroom and sanitation facilities (NSSO 2014).

Target D is to ensure that there is a significant improvement in the lives of not less than 100 million slum dwellers by 2020. As per NSS 2013, the number of urban-dwelling units situated in slums is equal to 10.8%; however, census report for the year 2011 states the urban households located in slums equal to 17.2% and there is a 37.14% of decadal growth in the number of slum households. India has the second-highest slum population (104 million) preceded only by China (180 million) (UNHABITAT 2014).

2.4.8 Goal 8: Fostering Partnership for Development

MDG8 is dedicated to fostering partnership among the member countries to achieve the prescribed goals. Members can learn from the expertise of each other and use it for human development in their own country (see Table 2.14).

Telecom services have been recognized as tools for socioeconomic development for a nation throughout the world. From the target and indicators of Goal 8, India has commonly adopted target F in cooperation with the private sector, including the benefits of new technologies, especially information and communications.

TRAI reported that, by the end of the year 2013, with 903.09 telephones connection including 873.36 million wireless connections, India is behind only China in the world ranking. There has been a sturdy growth in the telecom sector during the past

Table 2.14 MDG8 targets and indicators

Leverage the benefit of technological advancement	Fixed telephone services/100 persons
	Mobile cellular services/100 persons
	Internet user services/100 persons

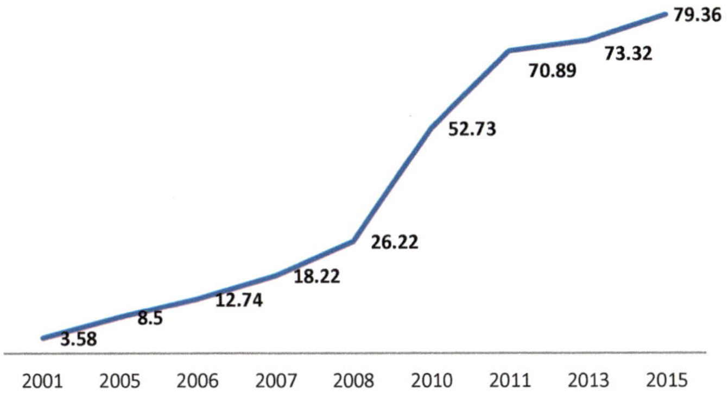

Fig. 2.7 Telephone user/100 persons, India. *Source* Telecom Stats (2018)

few years, a phenomenal change has been witnessed in mobile versus fixed phones and public versus private participation.

Figure 2.7 depicts the tremendous growth in the period between 2005 and 2015, with around 79.36% of people using in 2015 compared to 8.5 in 2008 (TRAI 2014).

2.5 MDGs Implementation

There has been a focus on effective public policy to meet MDGs; however, there is also a need to effectively implement public policy. It is important that countries not only develop economically, but also have sustainable development. In order to achieve sustainable development, it is important to develop effective institutions. Investing the money in building these institutions is important, but it is also important to have sound economic management and business regulation, build safeguards against corruption, and design measures to safeguard against discrimination against women and weaker sections. It is important to raise awareness through education and training. Effective implementation of MDGs can positively affect the development, health, and welfare of the people, particularly the weaker sections of society and those at risk.

2.6 MDGs Challenges

Millennium Development Goals (MDGs) endeavor to reduce poverty, improve hygiene, health, and education level, and solve sanitation problems. While the progress in education and health seems moderate, there is noticeable poverty reduction. It is important that the development strategies not only increase the growth rate but also the income share of the lowest section of the population to address the rural–urban divide (Likhi 2013). According to Desai (2013: PN), the major challenges of MDGs include lack of human rights framework, lack of engenderment, shortcomings in enabling environment, underfunding of gender equality, selective effects of global recession, etc. It has been stated that MDGs failed to consider a proper human rights framework. In the absence of such a framework, it is difficult to attain desired outcomes. For example, selective testing of women for HIV violates their right to freedom. It has also been seen that more women fall below the poverty line than men. Women face more resource crunch as compared to men. But MDGs fail to account for this. MDGs also fail to account for violence against women; sexual and reproductive health and rights; women, peace, and security issues. There is no guarantee that the achievement of MDGs would improve the conditions of marginalized and disadvantaged sections. The achievement of MDGs is facing another challenge in the lack of enabling environment for global economic policies. The states need to ensure that even in times of severe economic crisis, the states must ensure that basic rights are available to vulnerable sections of society. There is a need to review best practices and take lessons from them. There is also a dire need to allocate more funds to achieve gender equality. The political and economic empowerment of women and weaker sections of the society also require the allocation of more funds. The global recession has also affected women more than men. They seem to be more affected due to the lack of employment, wage cuts, reduced benefits, lower remittances, price rises, etc. There is also a move by the governments to cut social benefits like healthcare, unemployment benefits, etc.

According to Bello and Suleman (2011), MDGs have an inconsistent methodology and cut-off points for tracing progress, lack quality data on MDGs indicators, have ambiguous and unrealistic targets, and the baseline year of 1990 is biased against certain countries and regions, country reports are inconsistent, etc. There is a lack of a universally accepted method of the performance of MDGs in achieving various goals and targets. There are contradictions, counterclaims, and conflicting results in various studies. Another problem is the lack of classification criteria and cut-off points for grouping countries and regions. There is also a lack of timely, adequate, accurate, reliable, and consistent data. Several countries do not have the data for the base year of 1990 to derive performance targets for the target year 2015. According to Easterly (2009), MDG targets are not designed properly and are arbitrary and ambiguous for defining success and failure in achieving targets, such as poverty reduction. Clemens et al. (2004) noted that a typical African country has to grow by 7% each year for 25 years in order to achieve the goal of reducing poverty by half. So, certain goals and targets that are set are unattainable. Leo and Barmeier

(2010) noted that many of the MDGs measure progress in terms of development outcomes (like reducing maternal mortality) rather than development outputs (such as enrolling girl children in school). Achieving outcomes is a more difficult and tedious process than achieving outputs. Dickson (2010) alleged that MDGs do not take into account local capabilities and priorities. They focus on the short-term rather than the long-term. The millennium declaration in 2000 set the baseline year for measuring MDGs achievement as 1990. Due to this, some countries that progressed relatively more in the last decade got in a better position to achieve the goals as compared to others. As a necessary corollary, some countries with low performance in the previous decade were far behind in achieving MDG goals. The country reports that are prepared by the countries themselves or with the help of donors using consultants are uncoordinated and inconsistent. They are not in sync with the reports prepared by regional and international organizations. So long as countries use their reports as a publicity campaign, such reports cannot show their true performance. There is a need for a watchdog institution to approve such reports.

2.7 Conclusion

Millennium Development Goals (MDGs) have emerged as a platform to ameliorate the world's challenges. The major challenges faced by the world have been addressed in the form of 8 goals, 21 targets, and 43 indicators. Objectives that are measurable as well as time-bound, have been set to meet these MDG goals. MDGs create financial incentives, awareness among people and countries, measurement mechanism, accountability, and feedback mechanism to meet the world's challenges. It has been seen that developing countries have made considerable progress in achieving these goals by the target year 2015. However, there are areas of concern that need to be addressed in various countries. India has made significant progress in the achievement of MDGs. Both countries have made progress in areas like reducing poverty, reducing unemployment, human resource development, health, etc. Bihar is among the poorest and one of the most populous states in India. In Bihar, more than 50% of the population is below the poverty line and it accounts for high infant mortality. The progress of India in achieving MDGs can be a barometer of the progress of the economy (a prominent economy in Asia) in general.

There are 8 MDG goals set. Goal 1 aims to eradicate extreme poverty and hunger. This goal has targets of halving poverty between 1990 and 2015, achieving full and productive employment and decent work, and halving people suffering from hunger between 1990 and 2015. The incidence of poverty has fallen in the past 3 decades in India. However, India needs to improve at a faster rate to achieve the MDGs.

Goal 2 aims to achieve universal primary education. The target is to ensure that children achieve full primary schooling by the year 2015. In order to achieve this, India is providing free and compulsory education to all children. The net enrolment ratio has improved significantly in both countries. India has already achieved 100% enrolment.

Goal 3 aims to encourage gender equality and empower women; the objective has been to terminate the gender disparity in both primary as well as secondary education. It has been seen that the gender gaps have reduced in primary school education. The fraction of women in wage-earning in the nonagricultural sector is another important indicator. The progress has been slow on this front in India. Another important indicator under this goal is the percentage of women in the national parliament for which country has to work.

Goal 4 aims to reduce child mortality. Significant progress has been made in achieving this goal. Under-5 children mortality rate has been reduced significantly. India has done well in this regard. However, countries need to progress much more to achieve the 2015 target. Immunization against measles is another important indicator. India has been steadily moving to achieve this target.

Goal 5 aims to improve maternal health. The target aims at the reduction of the maternal mortality ratio by three-fourths by the year 2015, thereby achieving universal access to reproductive health by the year 2015. India has a high maternal mortality rate, and the country has to improve its health infrastructure. The other important indicator is the proportion of births undertaken by skilled health personnel; this has witnessed a significant improvement for India. Goal 6 aims to fight HIV/AIDS, malaria, and other diseases. The aspiration is to reverse the spread of HIV/AIDS by 2015. It is also targeted to achieve universal access to medical attention for all those who need it. It is also aimed to reverse the spread of incidence of malaria. The spread of HIV/AIDS is low and declining. The number of people who are getting HIV/AIDS treatment has increased in both countries. In the case of malaria control, significant progress has been made by India. Though in India, it is still a health burden and further steps are needed to control it.

Goal 7 aims to ensure environmental sustainability. Environmental sustainability is a prerequisite for sustainable development and poverty alleviation. The target is to promote sustainable development and reverse the loss of natural resources. It also aims to reduce biodiversity loss. It also targets to provide access to safe drinking water and basic sanitation and improve the life of slum dwellers. It has been seen that India has shown its commitment to sustainable development. India has also done well to control ODS. It is also important to work on increasing biodiversity. India has done well to provide access to safe drinking water and improve sanitation facilities.

Goal 8 aims to develop a global partnership for development. It targets to promote a just and equitable trading and financial system. It also intends to promote good governance and address the special needs of least developed countries. At the same time, it intends to deal comprehensively with debt control. It targets to provide access to affordable essential drugs. It also endeavors to leverage Information and Communication Technologies (ICTs). It has been seen that India has expanded well in providing telecommunication services. They are also moving swiftly to exploit technology in new and emerging sectors.

It can be said that MDGs have caused a significant decrease in poverty reduction, whereas progress in education and health seems moderate. The development strategies should not only focus on the growth rate, but also on equitable development. It is also important that MDGs should consider a proper human rights framework.

MDGs also need to account more convincingly for the challenges faced by women. The achievement of MDGs also needs to take into account the improvement in the status of disadvantaged and marginalized sections. The best practices need to be reviewed and implemented. Universally accepted methods of achieving MDGs need to be developed and adopted. MDGs also need to take into account local capabilities and priorities. There is also a need to develop a watchdog institution to oversee and approve the country's reports regarding the achievement of MDGs.

References

Adesina JO (2009) Social policy in Sub-Saharan Africa: a glance in the rear-view mirror. Int J Soc Welf 18(s1):S37–S51

Barrientos A (2013) Social assistance in developing countries. Cambridge University Press, Cambridge

Barrientos A, Hulme D (2009) Social protection for the poor and poorest in developing countries: reflections on a quiet revolution. Oxf Dev Stud 37(4):439–456

Barrientos A, Niño-Zarazúa M, Maitrot M (2010) Social assistance in developing countries database version 5.0. Brooks World Poverty Institute, University of Manchester

Bello A, Suleman A (2011) The challenge of achieving the millennium development goals in IDB Member Countries in the Post-Crisis World: IDB Occasional Paper No. 16. Islamic Development Bank, Jeddah

Bryce J, Arifeen S, Pariyo G, Lanata C, Gwatkin D, Habicht JP (2003) Multi-country evaluation of IMCI Study Group. Reducing child mortality: can public health deliver? Lancet. Retrieved from: https://doi.org/10.1016/S0140-6736(03)13870-6

Case A, Lubotsky D, Paxson C (2002) Economic status and health in childhood: the origins of the gradient. Am Econ Rev 92(5):1308–1334

Census (2011) Census of India 2011. Registrar General and Census Commissioner, New Delhi, India. Available on: https://censusindia.gov.in/census.website/

Clemens MA, Kenny CJ, Moss TJ (2004) The Trouble with the MDGs: confronting expectations of aid and development success—Working Paper 40. Centre for Global Development. https://www.cgdev.org/publication/trouble-mdgs-confronting-expectations-aid-and-development-success-working-paper-40

Dang H-A, Lanjouw P (2015) Poverty dynamics in India between 2004 and 2012. Insights from longitudinal analysis using synthetic panel data. World Bank Policy Research Working Paper No. 7270 (May). World Bank, Washington DC

Dang H-A, Lanjouw P (2018) Welfare dynamics in India over a quarter century. Poverty, vulnerability and mobility, 1987–2012. WIDER Working Paper 2018/175 (December). United Nations University-World Institute for Development Economics Research (UNU-WIDER), Helsinki

Dickson D (2010) MDGs: a blessing and a curse. Sci Dev Net, pp. https://www.scidev.net/global/editorials/mdgs-a-blessing-and-a-curse-1/

DLHS (2009) District Level Household and Facility Survey 2007–08. International Institute of population science. Mumbai. Retrieved from: http://rchiips.org/pdf/india_report_dlhs-3.pdf

Drèze J, Sen A (2013) An uncertain glory: India and its contradictions. Allen Lane, London

Economics Times (2019) Record 78 women MPs in new Lok Sabha. Retrieved from: https://economictimes.indiatimes.com//news/elections/lok-sabha/india/record-78-women-mps-in-new-lok-sabha/articleshow/69484777.cms?utm_source=contentofinterest&utm_medium=text&utm_campaign=cppst

Fukuda-Parr S (2004) Millennium development goals: why they matter. Global Governance, 395–402.

Ghosh R (2012) Child mortality in India: a complex situation. World J Pediat WJP 8(1):11–18. Retrieved from: https://doi.org/10.1007/s12519-012-0331-y
Government of India (2018) India in figure. Ministry of Statistics and Programme Implementation, Central Stats Office. Retrieved from: http://mospi.nic.in/sites/default/files/publication_reports/India_in_figures-2018_rev.pdf
Gundry S, Wright J et al (2004) A systematic review of health outcomes released to household water quality in developing countries. J Water Health 2:1–13
Hazarika, I. (2012). India at the crossroads of millennium development goals 4 and 5. Asia Pacific Journal of Public Health, 450–463
Haines A, Smith KR, Anderson D, Epstein PR, McMichael (2007) Policies for accelerating access to clean energy, improving health, advancing development and mitigating climate change. Lancet 370. Retrieved from: https://doi.org/10.1016/S0140-6736(07)61257-4
Harriss J (2013) Audacious reforms? India's new rights agenda: an introduction. Pac Aff 86(3):561–568
Hulme D, Wilkinson R (2012) The millennium development goals and beyond: global development after 2015. Routledge
Human Development Report (2020) The next frontier: human development and Anthropocene. United Nation. Retrieved from: http://hdr.undp.org/sites/default/files/hdr2020.pdf
IIPA (2017) National Family and Health Survey 2015–2016. IIPA. Retrieved from: https://dhsprogram.com/pubs/pdf/FR339/FR339.pdf
IMF (2000) Key features of IMF poverty reduction and growth facility (PRGF) supported programs: prepared by the Policy Development and Review Department. International Monetary Fund. Retrieved from: https://www.imf.org/external/np/prgf/2000/eng/key.htm
IMF (2021) The IMF and The World Bank. International Monetary Fund. Retrieved from: https://www.imf.org/en/About/Factsheets/Sheets/2016/07/27/15/31/IMF-World-Bank
Inter-Parliament Union (2021) Number of women in parliament. Archived. Retrieved from: http://archive.ipu.org/wmn-e/arc/classif010219.htm#1
Jayal NG (2013) Citizenship and its discontents: an Indian history. Harvard University Press, Cambridge
Kadiyala S, Barnett T (2004) AIDS in India: disaster in the making. Econ Polit Wkly, 1888–1892
Kumar A, Valecha N, Jain T, Dash AP (2007) Burden of Malaria in India: Retrospective and Prospective View. Am J Trop Med Hyg, 69–78
Kumar P, Singhal N (2020) Mapping neonatal and under-5 mortality in India. Lancet. https://doi.org/10.1016/S0140-6736(20)31050-3
Kakwani N (1980) On a class of poverty measures. Econometrica 48(2). Retrieved from: https://www.jstor.org/stable/pdf/1911106.pdf
Lahariya C, Paul VK (2010) Burden, differentials, and causes of child deaths in India. Indian J Pediat 77(11):1312–1321. Retrieved from: https://doi.org/10.1007/s12098-010-0185-z
Lahiry S (2014) India's Millennium Development Goals. World Affairs: The Journal of International Issues, 18(3):140–149
Leo B, Barmeier J (2010) Who Are the MDG trailblazers? a new MDG progress index. Centre for Growth and Development. Working Paper No. 222. https://www.cgdev.org/sites/default/files/1424377_file_Leo_MDG_Index_FINAL.pdf
McKerrow N, Mulaudzi M (2010) Child mortality in South Africa: using existing data. In: Fonn S, Padarath A (eds) South African health review 2010. Durban, Health Systems Trust
McKerrow N, Mulaudzi M (2021) Child mortality in South Africa using existing data. Retrieved from: https://www.researchgate.net/publication/267550766_Child_Mortality_in_South_afriCa_uSinG_EXiStinG_data
McMichael AJ, Friel S, Nyong A, Corvalan C (2008) Global environmental change and health: impact, inequalities and health sector. National Library of Medicine. Retrieved from: https://pubmed.ncbi.nlm.nih.gov/18219041/#affiliation-1

Midgley J (2004) Social development and social welfare: implications for social policy. In: Kennett P (ed) Social development and social welfare: implications for social policy: a handbook of comparative social policy. Edward Elgar, Cheltenham, pp 217–238

Mkandawire T (2004) Social policy in a development context: introduction. In: Mkandawire T (ed) Social policy in a development context. Palgrave Macmillan, London, pp 1–36

Manning R (2009) Using indicators to encourage development: lesson from millnium development goals . Danish Institute for International Studies, Copenhagen, DIIS

Martinez J, Mboup G, Sulizas R, Stein A (2006) Trends in urban and slum indicators across development world cities 1990–2003. Habitat International 32

Mohanty SK (2012) Multiple deprivations and maternal care in India. International Perspectives on Sexual and Reproductive Health. 38(1) 6–14

MOSPI (2010) Millennium development goals country report, New Delhi, India

MOSPI (2014) Millennium development goals country report, New Delhi, India

MOSPI (2017) Millennium development goals—Final Country Report of India. Ministry of Statistics and Programme Implementation, Government of India, New Delhi

Moyo S, Wright JA., Ndamba J, Gundry S (2004) Realising the maximum health benefits from water quality improvements in the home: a case from Zaka district, Zimbabwe. Physics and Chemistry of the Earth, 1295–1299

National Sample Survey Organisation (2013a) Key indicators of drinking, water, sanitation, hygiene and housing condition in India. Ministry of Statistics and Program Implementation, 76th Round. Retrieved from: http://mospi.nic.in/sites/default/files/NSS7612dws/Report_584_final.pdf

National Sample Survey Organisation (2013b) Key indicators of employment and unemployment in India 2011–12, 68th round. Retrieved from: http://mospi.nic.in/sites/default/files/publication_reports/KI-68th-E%26U-PDF.pdf

NSSO (2014) Participation of women in specified: activities along with domestic duties. NSS 68th Round. Ministry of Statistics and Programme Implementation. Government of India, New Delhi

National Family and Health Survey (2007)

NFHS-3 (2007) Key Findings From NFHS-3. IIPS, Mumbai. Available at: http://rchiips.org/nfhs/factsheet.shtml

Oduse S, Zewotir T, North D (2021) The impact of antenatal care on under-five mortality in Ethiopia: a difference-in-differences analysis. BMC Pregnancy Childbirth 21(1):44. Retrieved from: https://doi.org/10.1186/s12884-020-03531-5

Piot P, Greener R, Russell S (2007) Squaring the circle: AIDS, Poverty, and Human Development. PLoS Med 4(10):e314, 1571–1575

Poku N, Whitman J (2011) The millennium development goals: challenges, prospects and opportunities. Third world quarterly, 3–8

PIB (2021) Maternal Mortality Rate (MMR) reported to registral general of India by Sample Registration System. Government of India. Retrieved from: https://www.pib.gov.in/PressReleasePage.aspx?PRID=1697441

Press Bureau India (2021) Maternal Mortality Rate (MMR) reported to registral general of India by sample registration system. Government of India. Retrieved from: https://www.pib.gov.in/PressReleasePage.aspx?PRID=1697441

Ravallion M (2004) Pro-poor growth: a primer. Development Research Group. The World Bank Washington D.C., Retrieved from: https://doi.org/10.1596/1813-9450-3242

Radkar A, Parasuraman S (2007) Maternal deaths in India: an exploration. Econ Polit Wkly, 3259–3263

Ravallion M, Chen S (2003) Measuring pro-poor growth. Econ Lett. 93–99

Rehfuess E, Mehta S, Prüss-Ustün A (2006) Assessing household solid fuel use: multiple implications for the Millennium Development Goals. Environ Health Perspect 114:373–378. Retrieve from: https://www.ncbi.nlm.nih.gov/pmc/articles/PMC1392231/

RTE ACT (2009) The Gazette of India. Ministry of Law and Justice. Government of India. Retrieved from: https://www.education.gov.in/sites/upload_files/mhrd/files/upload_document/rte.pdf

Rodrik D (2008) Has globalization gone too far? Peterson Institute, Washington, DC

Sen A (1979) Equality of what? The tanner lectures on human values. Delivered at Stanford University. Retrieved from: https://www.ophi.org.uk/wp-content/uploads/Sen-1979_Equality-of-What.pdf

Shaw A (2013) Comparing NSSO employment survey—a methodological note. Econ Polit Wkly. Retrieved from: https://www.epw.in/journal/2013/30/web-exclusives/comparing-nssos-employment-surveys-methodological-note.html

Singh NK, Stern N (2013) The new Bihar: governance and development. Harper Collins Publishers, New York

Subramanian A (2013) India's troubling trilemma: the reason why macro stability is so elusive for India

Sumner A (2013) Who are the poor? New regional estimates of the composition of education and wealth "poverty" by spatial and social inequalities. ODI Working Paper No. 378 (April), Overseas Development Institute

The Hindu (2014, 17th July) India has 3rd-highest number of HIV-infected people. UN, New Delhi. Retrieved from: https://www.thehindu.com/sci-tech/health/india-has-3rdhighest-number-of-hivinfected-people-un/article6220483.ece

The Millennium Declaration (2000, September) The millennium development goals and targets come from the millennium declaration, signed by 189 countries, including 147 heads of State and Government, in September 2000. http://www.un.org/millennium1declarationJares552e.htm Accessed on 20 May 2015

TRAI (2014) Telecom sector in India: a decadal profile. TRAI, New Delhi

UN (2016) MDG Monitor: MDG 6: Combat HIV/AIDS, Malaria and other major diseases. United Nation, New York

United Nation (2017) The Millennium Development Goals report 2015. Retried from: https://www.undp.org/publications/millennium-development-goals-report-2015

UN (2000) United Nations Millennium Declaration. A/RES/55/2. United Nations, New York. Retrieve From: https://www.un.org/en/development/desa/population/migration/generalassembly/docs/globalcompact/A_RES_55_2.pdf

United Nations Children's Fund (2012) Committing to child survival: a promise renewed. New York. Retrieved from: https://www.unicef.org/reports/committing-child-survival-promise-renewed

UNAIDS (2021) Global HIV & AIDS statistics—fact sheet. Retrieve from: https://www.unaids.org/en/resources/fact-sheet

UNHABITAT (2014) World Habitat Day 2014—Voices from Slums. WHD 2014 Background Paper. United Nations, New York. Available at: https://unhabitat.org/sites/default/files/2014/07/World-Habitat-Day-2014-Backgrounder.pdf

Wilkinson R, Hulme D (2012) The millennium development goals and beyond: global development after 2015. Routledge

WHO (2014) Improved data reveals higher global burden of tuberculosis. World Health Organization, Geneva

World Health Organization (2015) Global Tuberculosis Report, 2015. Retrieved from: https://www.who.int/tb/publications/global_report/gtbr15_main_text.pdf

World Health Organization (2004) Maternal mortality in 2000: estimates developed by WHO, UNICEF and UNFPA. World Health Organization. Retrieved from: https://apps.who.int/iris/handle/10665/42930

World Bank (2014) World development indicator. World Bank Group, Washington DC

Wright C, Millward M (2005) Foreword. In UNESCO, & UNICEF, Children out of school: measuring exclusion from primary education (p. 3). Montreal: UNESCO. P Institute for Statistics Retrieve From: https://unesdoc.unesco.org/ark:/48223/pf000014386

Zahr C A, Royston E (1991) Maternal mortality: a global factbook. World Health Organization, Geneva. Retrieve From: WHO_MCH_MSM_91.3.pdf (7.797Mb)

Chapter 3
Appraisal of Human Development Through the Lens of Millennium Development Goals in Bihar

Abstract Bihar is a state that has its ancient glory hitting rock bottom, resulting in its isolation from the mainstream development of the country. In recent years, it has emerged as one of India's most vibrant and fastest developing economies. Despite this growth, Bihar has witnessed an exceptionally higher extent of human deprivation in global terms and the rest of the Indian states. Although the state expenditure in the social sector is increasing, Bihar stood at the bottom among other states of India. This resulted in minimal per capita development expenditure. The state needs huge investment in social infrastructure. Bihar urgently needs to prioritize the experience of MDGs.

Keywords Bihar · Growth · Development · Poverty · Public policies

3.1 Introduction

In BC 600, a prince, *Siddhartha,* from a country not far from Bihar, encountered human deprivation, disease, hunger, and abject poverty, which led him on a journey in quest of human development, the light of truth. He traveled long in search of answers, ultimately settling in Bihar, the erstwhile Magadh empire, where he achieved enlightenment and transform into Buddha, whose teaching continues to inspire various philosopher across the globe. After nearly 2500 years of Buddha, the same Bihar is struggling the meet the challenges of human development. It is the same land where the seeds of the first republic were sown and the first crop of democracy was cultivated. Bihar is "Karmabhumi" of great rulers, philosophers, and intellectuals. Lord Buddha and Lord Mahavira achieved enlightenment in the same land. Pataliputra was the capital city of the erstwhile Magadha empire, situated on the bank of the holy river Ganga. The state as it is today has been shaped from its partition from the province of Bengal in 1912, and most recently after the separation of the tribal southern region now called Jharkhand in 2000.

Bihar is a state with its ancient glory hitting rock bottom, resulting in its isolation from the mainstream development of the country. In recent years, it emerged as one of India's most vibrant and fastest developing economies. Despite this growth, Bihar has witnessed an exceptionally higher extent of human deprivation in global

U. K. Singh, *Human Development in Bihar, India*, Advances in Geographical and Environmental Sciences, https://doi.org/10.1007/978-981-19-3624-1_3

Table 3.1 Facts and figures of Bihar and India

	Bihar	India
Total area (km^2)	94,163	3,287,240
Total population (in crore)	10.38	121.01
Decadal growth (%)	25.07	17.64
Crude birth rate	27.60	21.4
Crude death rate	6.60	7
Natural growth rate	21	14.4
Infant mortality rate	42	40
Maternal mortality rate	210	178
Total fertility rate	3.5	2.4
Sex ratio	916	940
Child sex ratio	933	914

Source Census (2011)

terms than the rest of Indian states. Bihar is home to 104,099,452 persons living in a total geographical area of 94,163 km^2. The state is the most populous among the Indian states with an average density of 1102 person/km^2. The population spread over 38 districts, further in 44,874 villages. The state also stood number one in total fertility rate; the decadal population growth of Bihar is 25.07% which is higher than the national percentage of 17.64. In terms of sex ratio, Bihar is behind the national ratio but doing well in child sex ratio with 933/1000 while the national ratio is 914/1000 (see Table 3.1) (Census 2011). The state is more populous and more densely populated than some neighboring countries of its size.

Socio-economic and Caste Census (2011) figured that the total number of households (HH) in Bihar is 20,074,242 with 534 Tehsils, 9989 Gram Panchayats, 44,935 rural villages, and 138 towns. The percentage of households in rural areas is 88.82%, while 11.18% of Bihar falls in the urban area (see Table 3.2).

3.2 Growth Process and Development

The liberalization policy of 1991 transformed the Indian economy. The policy objective was Liberalization, Privatization, and Globalization (LPG); to make Indian companies more competitive in the global market and opportunities in the integrated global economy. In pre-liberalization period, Indian companies were neither encouraged nor prepared to compete with international beasts of the market, as the economy was controlled and regulated, and most of the business-related decisions were taken by the government. Post-1990s, liberalization appeared as an impetus for growth to uplift the growth rate in various sectors of the Indian economy. The growth performance at the national level was reflected by the increase in Net National

Table 3.2 Socio-economic and caste census

Districts	Total no. of H.H.	No. of tehshils	No. of gram panchayats/PS	No. of villages	No. of towns	Total no. of rural H.H.	Total no of urban H.H.	Rural H.H. (%age)	Urban H.H. (%a ge)
W. Champaran	751,752	18	499	1486	5	681,326	70,426	90.63	9.37
E. Champaran	990,770	27	406	1293	10	917,790	72,980	92.63	7.37
Sheohar	155,164	5	55	203	1	149,211	5953	96.16	3.84
Sitamarhi	759,803	17	274	837	5	725,723	34,080	95.15	4.49
Madhubani	915,609	21	1026	1112	4	886,810	28,799	96.85	3.15
Supaul	476,558	11	180	551	3	455,280	21,278	95.54	4.46
Araria	632,520	9	656	742	3	595,834	36,686	94.20	5.80
Kishanganj	339,550	7	126	771	3	308,989	30,561	91.00	9.00
Purnia	657,203	14	240	1273	3	591,936	65,267	90.07	9.93
Katihar	632,565	16	238	1541	2	582,089	50,476	92.02	7.98
Madhepura	420,846	13	271	440	2	402,414	18,432	95.62	4.38
Sahrasa	380,069	10	152	468	1	351,154	28,915	92.39	7.61
Darbhanga	887,990	18	313	1249	2	817,164	70,826	92.02	7.98
Muzzafarpur	951,487	16	399	1791	4	876,341	75,146	92.10	7.90
Gopalganj	413,792	14	234	1535	4	389,899	23,893	94.23	5.77
Siwan	546,331	19	296	1530	3	519,048	27,283	95.01	4.99
Saran	629,430	20	326	1765	5	577,746	51,684	91.79	8.21
Vaishali	669,465	16	291	1569	3	629,950	39,515	94.10	5.90
Samstipur	853,282	20	384	1250	3	831,029	22,254	97.39	2.61
Begusarai	591,769	18	383	1144	6	491,437	100,332	83.05	16.95
Khagaria	364,684	7	120	301	2	347,924	16,760	95.4	4.60

(continued)

Table 3.2 (continued)

Districts	Total no. of H.H.	No. of tehshils	No. of gram panchayats/PS	No. of villages	No. of towns	Total no. of rural H.H.	Total no of urban H.H.	Rural H.H. (%age)	Urban H.H. (%a ge)
Bhagalpur	562,048	16	256	1521	4	468,355	93,693	83.33	16.67
Banka	398,033	11	185	2111	2	384,455	13,578	96.59	3.41
Munger	267,487	9	110	862	3	201,233	66,254	75.23	24.77
Lakhisarai	164,862	7	81	472	2	141,284	23,578	85.70	14.30
Sheikhpura	111,407	6	55	314	2	91,847	19,560	82.44	17.56
Nalanda	549,242	20	250	1056	5	462,838	86,404	84.27	15.73
Patna	135,1505	23	326	1395	15	690,850	660,850	51.12	48.88
Bhojpur	491,458	14	228	1217	6	419,641	71,817	85.39	14.61
Buxar	262,794	11	142	1134	2	238,210	24,584	90.65	9.35
Kaimur	250,722	11	1156	1697	1	241,516	9206	96.33	3.67
Rohtas	476,401	19	250	2076	6	409,854	66,547	86.03	13.97
Aurangabad	401,585	11	205	1849	4	366,898	34,687	91.36	8.64
Gaya	758,837	24	453	2888	4	659,015	99,822	86.85	13.15
Nawada	376,741	14	198	1086	3	349,738	27,003	92.83	7.17
Jamui	315,113	10	167	1506	2	292,670	22,443	92.88	1.12
Jehanabad	189,857	7	93	584	2	164,924	24,933	86.87	13.13
Arwal	125,510	5	65	316	1	116,644	8866	92.94	7.06
Total	20,074,242	534	9989	44,935	138	1,782,906	2,245,176	88.82	11.18

Source SECC (2011)

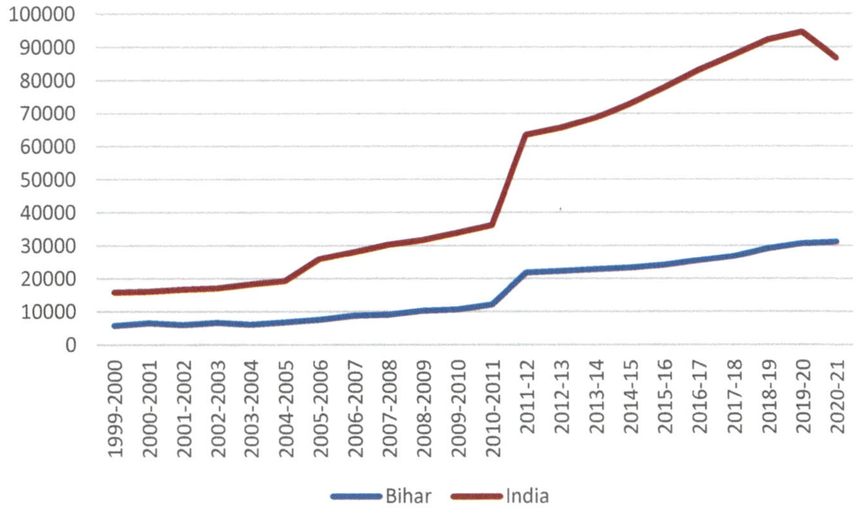

Fig. 3.1 Per capita state domestic product of Bihar and NNI (at constant price) (INR). *Source* RBI (2021)

Income (NNI) (at constant price), though the per capita state domestic of Bihar did not increase on par with the national average (see Fig. 3.1).

India has witnessed an impressive growth in per capita NNI during 2000–01 to 2015–16, from Rs 16,173 to Rs 77,659. The country has witnessed an almost fivefold increment in NNI but for Bihar the line of per capita GSDP is a straight line in the period of 2000–01 till 2009–10 and saw a boom from 2010 to 11 and from there on it followed an upward trend. Despite the positive movement, the pace is not at par with the national average.

The growth in per capita income of Bihar after 2010 resulted in an increase in investment in social infrastructure during 2010–11 and 2014–15, the increment doubled from Rs. 22, 926 to Rs 46,158. In 2014–15, Bihar took drastic steps to develop the social sector and the state spent 28% of the total state capital outlay (Rs. 18,150 crores) on creating social infrastructures such as highways and bridges; 19% (Rs. 315 crores) invested in health infrastructure; 53% (Rs. 885 crores) financed for sanitation and hygiene and 16% disbursed to the education sector, nearly Rs 14,728 crores spent on economic services (Government of Bihar 2016).

Improvement in social infrastructure translated into an improvement in growth rate, state gross domestic product expands annually at the rate of 10.5% annually at a constant price. Consequently, the development expenditure of the state also improved. The government focused on two main components of human development, health and education, and increased the per capita development expenditure which was below the national average of all states from 17.8 to 29.6% (Government of Bihar 2017). Though the growth is positive, below the national average (see Fig. 3.2). Unfortunately, with such a startling growth rate, Bihar remains India's poorest state

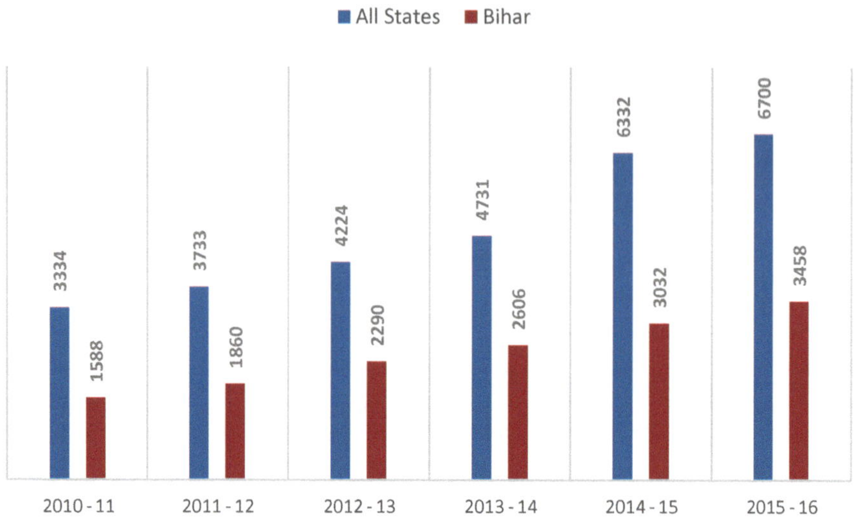

Fig. 3.2 Per capita expenditure on social capital in Bihar and All States of India. *Source*)(Government of Bihar 2017)

throughout the last decade with continuation in the current period, and deprivation sustained in the future.

The indicators of human development health and education remain in shamble. Although the public state had experienced an improvement in health and education, there are significant shortfalls in the number of primary health centers and sub-centers, as well as staff including auxiliary nurse midwives and staff for integrated child development services. Table 3.3 reflects the status of health infrastructure in Bihar and the shortfall in every sector is visible. Subcenters are important centers in rural areas where people are going for first-hand treatment or at the initial stage of health and play an important role in the implementation of various Health & Family Welfare schemes at the grass-root level. Subcenter focuses primarily on the Reproductive and Child Health (RCH) services. However, services in respect of important noncommunicable Diseases have also been included. The aim of the subcenter is to provide quality services that are of optimum level, fair, and responsive to the client's needs, provide equitably, and deliver improvement in the health and well-being of the population, and in Bihar 18,533 subcenters are required while 9696 are in position with the deficit of 8837 (Registrar General of India 2011).

In the education sector, the institutes of higher education are in shambles. The universities in Bihar were and are notorious for taking five to six years for the competition of 3-year degree courses. The number of existing professional institutes is not sufficient to equip the youth population with skilled and technical education. Bihar stood in the first position in the number of unemployed graduates in India. The 68th round NSSO finding suggests that among the major states of India, the proportion of students attending the public institution is highest in rural Bihar (95.7%) with 74%,

Table 3.3 Health infrastructure of Bihar

District	Sanction post (contractual)	No. of ASHA workers			ASHA/lakh population
		2012–13	2013–14	2014–15	
Patna	5842	3182	3018	3018	52
Nalanda	2880	2356	2313	2316	80
Bhojpur	2731	2264	2199	2188	80
Buxar	1708	1487	1474	1474	86
Rohtas	2962	2435	2430	2454	83
Kaimur	1628	1462	1462	1462	90
Gaya	4395	3493	3442	3448	79
Jehanabad	1126	871	870	870	77
Arwal	773	773	746	1956	107
Nawad	2221	1920	1928	2142	88
Aurangabad	2542	2158	2142	3402	84
Saran	3935	3422	3385	2822	86
Siwan	5099	2939	834	2374	85
Gopalganj	4805	2297	2336	3108	93
W. Champaran	3434	3064	3043	4060	79
E. Champaran	656	3762	3906	3858	80
Muzaffarpur	3498	3812	3856	2919	80
Sitamarhi	3941	2827	2912	572	85
Sheohar	4487	570	572	572	87
Vaishali	4266	2856	2888	3019	86
Darbhanga	3941	3546	3100	3118	79
Madhubani	4487	3632	3682	3852	86
Samastipur	4266	3829	3779	3794	89
Begusarai	2973	2361	2410	2403	81
Munger	1369	956	956	952	70
Sheikhpura	637	476	476	476	75
Lakhisarai	1002	756	756	901	90
Jamui	1762	1513	1504	1505	85
Khagaria	1668	1409	1399	1468	88
Bhagalpur	3040	2236	2236	2232	73
Banka	2037	1814	1782	1809	89
Saharsa	1902	1472	1471	1471	77
Supaul	2229	1920	1912	2093	94
Madhepura	2004	1638	1655	1674	84
Purenia	3268	2696	2634	2709	83

(continued)

Table 3.3 (continued)

District	Sanction post (contractual)	No. of ASHA workers			ASHA/lakh population
		2012–13	2013–14	2014–15	
Kishanganj	1690	1334	1280	1290	76
Araria	2812	2362	2365	2365	84
Katihar	3074	2547	2549	2720	89
Bihar	**104,239**	**4447**	**83,702**	**85,045**	**82**

Source Ministry of Health and Family Welfare, GoI

the state also leads in the enrolment in Government and Local body institution in urban areas. This reflects the overburdened situation of public institutions and the lack of private players in the education sector of Bihar (NSSO 2015).

In absence of sound health infrastructure and well-regulated educational institutions, Bihar failed in creating human capital which resulted in the deprivation of humanity in the state. MDGs envisaged concrete targets to improve the life of people living around the globe. For Bihar, the progress was modest toward the goals except for health and literacy. The state literacy rate remained the lowest among the other states of India; in per capita income state trails by threefold in comparison to the national average. The condition of health infrastructure was a bigger challenge for overall development, the state faced the highest incidence of infant and maternal mortality in the country.

After health and education, the nature of employment in Bihar is like any developing economy. 90% of Bihar's population lived in rural areas engaged in primary activities. The society of Bihar is an agrarian society that employed 81% of the total workforce of the state. The agriculture sector is the largest contributor to Bihar's economy, but the saddest part of this sector is that, despite the largest sector which provides employment, the growth remains stagnant in the agriculture sector. To realize human development, the state must increase investment in agriculture to modernize the sector.

In self-employment, registering a new venture in Bihar was a big challenge. World Bank (2009) report on ease of doing business compared business regulation for small- and medium-sized domestic enterprises across 17 Indian cities. Patna, the state capital of Bihar, and by far its richest district was ranked 14 out of 17 Indian cities in terms of its ease of doing business. It took 185 days in Patna to deal with the construction permit, 87 days to register a property and 9.3 yearss to close a business. In Ludhiana, India's city with the highest "doing business" rank, business required 143 days to deal with a construction permit, 67 days to register property, and 7.3 to close a business. Bihar is the classic example of the vicious circle of poverty and development, the central government too must share the responsibility of meeting these challenges (Krishnan 2018).

3.3 Program and Policies Framed to Achieve MDG in Bihar

The policy-makers across the world are working together to uplift the marginalized, poor, and vulnerable from deprivation of resources and enable them to actively participate in economic progress. The seventh schedule (under article 246) of the Indian constitution made the arrangement for power distribution among the Union and state of India. Subjects are classified into three lists, Union list, State list, and Concurrent list. The Parliament of India is responsible to make laws given in the subject of the union list, states are allowed to make laws falling under the preview of the state list while both of parliament and state legislature, they enable to make laws according to subjects given in the concurrent list, provided with federal supremacy to the national parliament.

Since the political transformation of 2005, Bihar realized international support for the development of social infrastructure. The betterment of law and order promotes investment in the state and attracted the attention of international agencies for support and capacity development. Bihar has introduced various programs and policies focused on developing infrastructure to create human capital by fostering investment in health, education, transportation, and other physical infrastructure. The aim of these programs was economic empowerment, which would, in turn, lead to human development.

Economic empowerment is realizing the capacity of men and women to participate, contribute, and benefit from the growth process. It is important that the value of the contribution is recognized by the state, and facilitate fairer distribution of growth benefit among the participants. Economic empowerment leverages sustainable development by creating job opportunities, increasing access to economic resources, enhancing health and educational services and connecting the poor and marginalized with the mainstream economy.

International donors, for instance, Department for International Development (DFID), United Kingdom, have made large contributions to control child and maternal deaths and improve institutional deliveries in the state by investing in health infrastructure (UNICEF 2020a, b). Notwithstanding the development in the health sector, the poor and rural populations are unable to access public health services. Illiteracy among the poor cohorts makes the situation worse. Pregnant and lactating women are not aware of the nutrient requirement of their bodies. The gap in infrastructure and human resources presents an obstacle to delivering health services.

Collaborative effort of the state government with the help of the union government and the support of international agencies are visible in programs and policies implemented by the state government in accordance with the goals and targets of MDGs (Table 3.4).

Table 3.4 Social welfare scheme of union government and government of Bihar

Central sponsored schemes	State-sponsored scheme
Mahatma Gandhi National Rural Employment Guarantee Scheme	Jeevika
National Social Assistance Programme	Mukhyamantri Balika Poshak Yojana
Programme for Development of Minorities	Mukhyamantri Balika Cycle Yojana
Programme for Development of Vulnerable Groups	Mukhya Mantri Balika Protshahan Yojana
Programme for Development of Scheduled Tribes	Kasturba Gandhi Vidyalaya
Programme for Development of Scheduled Caste	Akshar Anchal Yojana
Rashtriya Swasthya Bima Yojana (RSBY)	Hunar and Auzar
Blue Revolution	Janani Bal Suraksha Yojana
Border Area Development Programme	Mukhyamantri Kanya Vivah Yojana
Environment, Forestry, and Wildlife	COMFED- Sudha White Revolution
Green Revolution	Mukyamantri Gramin Sadak Yojana
Infrastructure Facilities for Judiciary	Hamara Adhara Urja Swit Bihar
National Rural Drinking Water Mission/ Jal Jeevan Mission (JJM)	
Jobs and Skill Development	
Mission for protection and Empowerment of Women	
Modernization of Police force	
National Education Mission	
National Livelihood Mission (Ajeevika)	
National Programme of Mid-Day Meal in School	
Pradhan Mantri Awas Yojana	
Pradhan Mantri Gramin Sadak Yojana	
Pradhan Mantri Krishi Sichai Yojana	
Rashtriya Gram Swaraj Abhiyan (RGSA)	
Shyama Prasad Mukharjee Rurban Mission	
Swachh Bharat Mission (Gramin)	
ICDS	
White Revolution	
Rashtriya Pashudhan Vikas Yojana	
Saksham Anganwadi and Poshan	
Mission Vatsalya	
Mission Shakti	
PM Formalization of Micro Food Processing Scheme	

3.4 Convergence of Millennium Development Golas (MDGs) and Public Policies in Bihar

3.4.1 MDG1

Although India has lifted more than 100 million people out of poverty, the country is home to a large number of people living in abject poverty. Government and policy-makers use the poverty headcount ratio (PHCR) to estimate the number of poor in the confined area. PHCR is derived by considering the income and consumption of the people below the declared threshold which is known as poverty line, set up by the state union government. For India, the erstwhile Planning Commission with the help of the National Sample Survey Office (NSSO), released a periodical for the poverty estimation in the country. The method of poverty estimation is dynamic, reviewed, and changed with time (Pangariya and Mukim 2014). The high growth of per capita state GDP during 2004–2005 to 2009–2010 helped Bihar to reduce the poverty headcount (Anand et al. 2014). Bihar along with Uttar Pradesh observed a sharp decline in the poverty headcount ratio. The MDG target 1 stipulates that the percentage of people below the national poverty line be brought down to half of its 1990 level. The Bihar PHCR estimate (total) was 60.5% (see Fig. 3.3) in the 1990s, and the poverty estimate shows that the total PHCR at the Bihar level is 33.74% in 2011–12 Notwithstanding the high state growth rate, Bihar could not accelerate the poverty reduction to meet the target (Poalnki 2012).

Poverty is a complex phenomenon. Sen (1999a, b) defined poverty as the failure to avail of necessities. Further, he conceptualized poverty and development in a

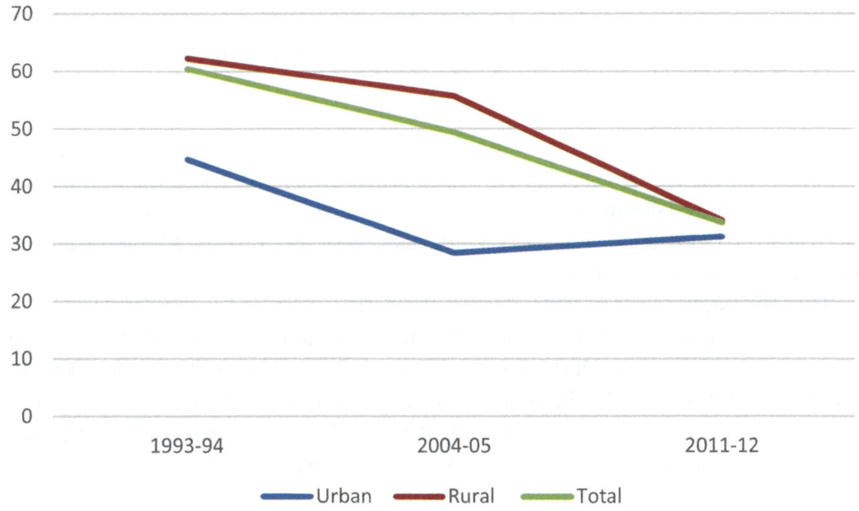

Fig. 3.3 Poverty Head Count Ratio, Bihar. *Source* MOSPI (2015)

framework, i.e., human development. Income plays a significant role in influencing an individual's ability to consume certain basic goods (Desjardins et al. 2006). The parameters of poverty comprise not just income but access to education, health, employment, housing, personal security, etc. as well (UNDP 1997). These aspects of poverty are multidimensional and affect the person facing it in its absolute gravity. But the strength of such dimensions varies in different spaces and times. Therefore, a person may be declared a nonpoor if going by the established income-based measure even though the subject may be facing deprivation in the remaining dimensions. A person lacking access to fundamental services and personal security cannot be considered free of poverty (Benassy-Quere et al. 2010).

The Planning Commission of India (1951) stated that growth is not the sole objective of economic policy, it is necessary to ensure that the benefits of growth are shared by the whole cohort of the society. Thus, eradication of poverty is becoming an important objective. Human beings need a certain minimum consumption of food and nonfood items to survive. However, the perception regarding what constitutes poverty varies over time and across countries. Nevertheless, there is a need for a measure of poverty.

3.5 Poverty Measures in India

Sen (1976) showed that the income-poverty measure faces a challenge when price and preferences vary. Price changes with location and low-income households may be doing better than others but have less purchasing power (Roberts 1980). Samuelson and Swami (1974) argued that the measure of price deflates income changes in accordance with utility change. The development of India is uneven and the country adopted a separate poverty line built on the dynamic consumption pattern of different states (Planning Commission, June 2014). India has a long history of poverty measurement studies.

The methodology of Poverty Estimation given by the Planning Commission in 1962 has been under constant discussion and debate by experts. Responding to such concerns, a task Force/Expert group was constituted by the Commission to review the methodology periodically. For example, the task force under Chairman Dr. Y. K. Alagh in 1977, under Prof. DT Latdawala 1989 and Prof. SD Tendulkar 2005 (Planning Commission2014).

Poverty estimation for 1993–94 and 2004–05 was based on the method of an expert group headed by Tendulkar. The same methodology was used in calculating the proportion of poor in 2009–10 and 2011–12. The report of 2004–05, 2009–10, and 2011–12 shows the declining trends in the poverty ratio of Bihar and India during different periods which is given in Table 3.5. In 2004–05, the proportion of people living below the poverty line in Bihar was 54.4% (Rural 55.7 and Urban 43.7%), which took a narrow slide in 2009–10 coming to 53.5% (Rural 55.3 and Urban 39.4%) and in 2011–12 percentage of poor in states slides smoothly to 33.4% (Rural

Table 3.5 Poverty incidence, Bihar and India

	Rural		Urban		Total	
	Proportion	No. of persons (in lakh)	proportion	No. of persons (in lakh)	Proportion	No. of persons (in lakh)
2004–05						
Bihar	55.7	445.1	43.7	40.9	54.4	485.6
India	41.8	3266	25.7	807.6	37.2	4076.1
2009–10						
Bihar	55.3	498.7	39.4	44.8	53.5	543.5
India	33.8	2782.1	20.9	764.7	29.8	3546.8
2011–12						
Bihar	34.1	320.4	31.2	37.8	33.7	358.2
India	25.7	2166.6	13.7	531.2	21.9	2697.8

Source MOSPI (2017a, b), Planning Commission (2014)

34.1 and Urban 31.2) percent. In the same period of time, the national poverty level came down to 21.9% from 37.2% in 2004–05 (Planning Commission 2014).

The Poverty computation by Prof. Tendulkar in 2004 on PPP terms came out to be Rs 33 per day. Considering this finding, the Planning Commission published the number of absolute poor to have reduced from 40.7 crores (2004–05) to 35.5 crores (2009–10) and 26.9 crores (2011–12). Dr. Rangarajan, the Chairman of PM's Economic Advisory Council resubmitted a report in 2014 to the Union Planning Minister Rao Inderjit Singh, which suggested that people spend less than Rs 47 per day in urban areas and Rs 32 in rural villages can be considered poor (Gaur and Rao 2020).

The Tendulkar Poverty Line report of 2004 has been backed by a panel tasked with finding ways to mitigate poverty on the ground that the line of Rs 33 per day is meant to be an indicator for tracing progress in fighting grave poverty. The line is not meant to be an identification for the beneficiaries of government distribution.

In 2015, NITI Ayog instituted a task force to eliminate poverty in India. The Taskforce suggested that the state government identify the poor through the BPL census instead of the poverty line, the latest data can be used by Socio-Economic Caste Census (SECC) of 2011. SECC facilitate the ranking in the rural and urban area, derived from existing parameters. SECC is based on three components, surveyed by three separate agencies, coordinated by the Department of Rural Development, Government of India. Survey of the rural areas was tasked to the Department of Rural Development while the Ministry of Housing and Urban Poverty alleviation was responsible for surveys in the urban area.

The 2011 Census was an exercise of comprehensive enumeration across the nation, both rural and urban parts. Census serves the purpose of ranking households based on socioeconomic status which enables state governments to list families living below

Table 3.6 Distribution of population as per their earning

Districts	Total households (rural)	Households having monthly income of highest earning member			Households having monthly income of highest earning member (%age) w.r.t. total HH		
		<5000	5000–10,000	>10,000	<5000	5000–10,000	>10,000
W.Champaran	681,326	492,902	147,704	39,468	72.34	21.68	5.79
E. Champaran	917,790	674,056	188,750	54,912	73.44	20.57	5.98
Sheohar	149,211	113,793	28,225	7193	76.26	18.92	4.28
Sitamarhi	725,723	497,002	183,979	44,680	68.48	25.35	6.16
Madhubani	8,868,810	628,140	204,820	53,561	70.83	23.10	6.04
Supaul	455,280	352,301	83,943	18,711	77.38	18.44	4.11
Araria	595,834	492,839	84,681	18,310	82.71	14.21	3.07
Kishanganj	308,989	254,423	47,198	7368	82.34	15.27	2.38
Purnia	591,936	461,903	108,784	21,236	78.03	18.38	3.59
Katihar	582,089	450,473	107,214	24,392	77.39	18.42	4.19
Madhepura	402,414	317,094	69,023	16,139	78.80	17.15	4.01
Sahrasa	351,154	273,611	60,975	16,467	77.92	17.36	4.69
Darbhanga	817,164	610,853	164,893	41,407	74.75	20.18	5.07
Muzzafarpur	876,341	624,766	192,274	57,721	71.29	21.94	6.59
Gopalganj	389,899	243,600	107,454	38,826	62.48	27.56	9.96
Siwan	519,048	295,444	158,098	65,501	56.92	30.46	12.62
Saran	577,746	351,364	169,284	57,057	60.82	29.30	9.88
Vaishali	629,950	453,949	128,397	47,574	72.06	20.38	7.55
Samstipur	831,029	574,273	201,997	54,653	69.10	24.31	6.58
Begusarai	491,437	349,739	110,379	31,317	71.17	22.46	6.37
Khagaria	347,924	245,558	83,382	18,955	70.58	23.97	5.45
Bhagalpur	468,355	317,870	114,311	35,076	67.87	24.41	7.49
Banka	384,455	308,133	57,080	19,242	80.15	14.85	5.01
Munger	201,233	146,942	34,562	19,729	73.02	17.18	9.80
Lakhisarai	141,284	99,427	32,582	9275	70.37	23.06	6.56
Sheikhpura	91,847	63,669	20,997	7180	69.32	22.86	7.82
Nalanda	462,838	302,009	118,449	42,379	65.25	25.59	9.16
Patna	690,850	445,810	194,304	50,733	64.53	28.13	7.34
Bhojpur	419,641	276,278	102,017	41,316	65.84	24.31	9.58
Buxar	238,210	141,359	71,501	25,344	59.34	30.02	10.64
Kaimur	241,516	160,917	58,058	22,541	66.63	24.04	9.33

(continued)

Table 3.6 (continued)

Districts	Total households (rural)	Households having monthly income of highest earning member			Households having monthly income of highest earning member (%age) w.r.t. total HH		
		<5000	5000–10,000	>10,000	<5000	5000–10,000	>10,000
Rohtas	409,854	276,155	91,840	41,602	67.38	22.41	10.15
Aurangabad	366,898	240,335	92,795	33,764	65.50	25.29	9.2
Gaya	659,015	459,480	153,137	46,355	69.72	23.24	7.03
Nawada	349,738	256,105	71,510	22,093	73.23	20.45	6.32
Jamui	292,670	220,006	56,272	16,247	75.17	19.23	5.55
Jehanabad	164,924	116,618	31,370	16,924	70.71	19.02	10.26
Arwal	116,644	78,073	27,475	10,933	66.93	23.55	9.37
Total	17,829,066		3,959,714	1,196,190	71.05	22.21	6.71

Source SECC (2011)

the poverty line. Census also helps identify particular sections of the population and castes on the basis of their education status and socioeconomic conditions.

As per the SECC (2011), the total number of households (HH) in Bihar is 20,074,242 and the total number of HH surveyed during the caste census is 1,782,906. The outcome of this census is reflected in Table 3.5, where it is depicted that 71.05% of the population live below the monthly income of Rs.5000, 22.21% earns between 5000 and 1000 while only 6.71% of people have a monthly income of more than Rs. 10,000. The situation is alarming for the government and the key policy-makers are working to raise the income of this big chunk of the population living below the poverty line and able to avail the basic amenities of life, through various government schemes and policies. The major concern for the government is the marginalized section of the society. As per data provided by the Ministry of Social Justice, Government of India, the condition of scheduled caste is alarming in rural areas of the state where 64% of the total SCs population, 37.8% of OBCs, and 26.6 of other groups are living below the poverty line, the corresponding figure for urban areas is 67.2, 41.4, and 18.3% (Government of India 2016).

3.6 Pro-poor Policy in Bihar

Empowerment is a multifaceted, multidimensional, and multilayered concept. The Government of India publishes MDG country, which defined empowerment as the means of moving from a position of enforced powerlessness to one of power. Economic Empowerment is the key to opening up avenues of such power by enabling marginalized sections to gain personal identity and social status in the society.

The NSSO 66th Round Survey, July 2009–June 2010 on employment and unemployment in India revealed that the status of worker population ratio (WPR) for rural males was 547/1000 and for rural females was 261/1000. For urban areas, the corresponding figures were 543 and 138 respectively. Among the workers in the rural areas, about 54.2% were "self-employed", about 38.6% were "casual labor" and 7.3% were "regular wage/salaried" employee (NSSO 2011). The Central Government formulated the National Rural Employment Guarantee Act (MGNREGA) in 2005, a wage employment program to fight more effectively with poverty.

MNREGA steps in to fill the gaps of challenges facing such sections with its legal and rights-based approach. MNREGA is a flagship program of GOI aimed at improving livelihood security by providing 100 days of guaranteed wage employment in a financial year to every rural household whose adult members volunteer to do unskilled manual work. The scheme covered 200 districts in the first phase, when first implemented on February 2, 2006, with 130 districts added in 2007–2008. The remaining rural areas have been notified from April 1, 2008 (Government of India 2022) .

3.6.1 The Mahatma Gandhi National Rural Employment Guarantee Scheme (MNREGA)

The contemporary idea of development is to promote sustainable development by enhancing capabilities and many developing countries have implemented innovative instruments of development. Mahatma Gandhi National Rural Employment Guarantee Scheme is one of them (Panda 2015). Economist Bhalla (2013) termed MGREGA as "dole". The recent global human development report (HDR 2015) refers to MGNREGA as one of the milestones in social protection measures in the world. Since private investment is lower in rural areas, Shah (2007) argued that the lack of basic infrastructure tends to lower return on investment in backward areas. The development economists termed this phenomenon as "low-level equilibrium trap" (Nelson 1956). Thus, a "big leap" is needed to take the economy out of this situation. A "big leap" referred to increase in government spending on basic infrastructure, which accelerate the positive return in rural area. MGNREGA is such a social assistance programme with comparable cohort schemes—Rural Employment for Public Assets such as *Food for Work* programme in Bangladesh and *Jefes y jefas de Hogar Desocupados* in Argentina (Ravallion et al. 1991).

MGNREGA leverages economic empowerment throughout the journey of MDGS. The target of the MDG coincided with the decade completed by MGNREGA. The programme generated 1970 crore person-days employment opportunities; 40% job occupied by SC/ST and women accounted for 57%. An estimated 15.54 lakhs of households have attained 100 days of job, further 26.19 lakhs households completed 81–91 days.

Table 3.7 depicts the progress made through MNREGA till January 2016. The total employment provided by this program is 20.34,996 lakhs in which 28.36%

Table 3.7 Job provided under MGNREGA in the past 10 years

Employment provided to households	20.34996
Scheduled Caste (SC)	241.18
Scheduled Tribe (ST)	17.92
Women	297.1
Others	591.21
Total works taken up	576,208
Work completed	88,157
Work in progress	488,051
Total	850.31

Source MGNREGA Statistics (2016)

belongs to scheduled caste, 2.11% belongs to scheduled tribe and the percentage of women workers is 34.94% (Ministry of Rural Development 2016).

The MGNREGA has generated more rural employment than any other government scheme or private initiative in the history of independent India. Desai et al (2015) figured that MGNREGA has the potential to reduce poverty and empower women but success depends on the reach of the programme. Study revealed that barely 11% of the poor households participate in Bihar due to the lower capacity of local administration. The study registered that 25% decline in poverty since 2004–05 for participating households could be attributed to the participation in the rural employment scheme. These households are less likely to borrow from moneylenders and more likely to give higher education to their children.

The proportion of women in the labour force participation is continuously decreasing. In the era of mechanization and automation, women in rural areas are left with very few paid job opportunities (Mehrotra and Sinha 2017). Here, MGNREG emerges as a key changer by benefiting men and women with wage equality (Mattos and Dasgupta 2017). The positive impact of MGNREGA is reflected by gender parity; for the first-time women are being paid at par with men in a job, and where women outnumber men in the job scheme. As many of the women got work for the first time due to the job scheme, their household income levels also went up. The MGNREGA has also boosted financial inclusion through access to a bank account by women as the wages are transferred electronically to the beneficiary's account.

Ravindranath and Murthy (2021) studies suggest that the MGNREGA programme work for environment safeguard as well as increasing income of rural population, with over USD 7 Billion budget allocation annually. The programme focused on natural resources such as water conservation, water harvesting, drought proofing, flood control, rejuvenating natural water source, irrigation, and rural connectivity. It improved ground water levels, increased water availability for irrigation purpose, expansion of the land under irrigation by ground and surface water sources and improved accessibility of drinking water for human purpose and livestock (Esteves et al. 2013).

MGNREGA has been a critical source of income for female-headed households, providing as much as 15% of the household income in some states. Plus, it has given a fillip to rural entrepreneurship, with households using the supplementary income to start a rural business. The MGNREGA has had a spill over impact of rural economy, with the increased purchasing power, consumption pattern changes from staple food to diverse diet, resulting in improved health and nutrition and arrest malnutrition (IDR 2013). The demand-driven programme help in inclusive development since the proportion of marginalized communities such as SC and ST are more than general beneficiaries.

3.6.2 *Deen Dayal Antoyoday Yojana (DAY-NRLM)*

The theme of Antyodaya is based on the principal Mahatma Gandhi's belief of Sarvodaya (wellbeing of all) through *antoydaya* (well-being of poor), the development of all through the welfare of the marginalized (Naidu 2018), the idea was expressed by John Ruskin (1862). The concept of Antyodaya is to take welfare to the poorest of the poor and livelihood promotion is the prime objective of the scheme, focusing on the enhancement of employment in rural economy by harmonizing agriculture and allied sector, small enterprises.

The policy ensures the participation of vulnerable and marginalized cohorts of the rural society through reservation. 50% share goes to SC/ST applicants and 15% to minority and 3% dedicated to disable persons.

The uniqueness of the programme lies in bridging the gap between labor demand and supply. DAY-NRLM through its subcomponent, Mahila Kisan Sasaktikaran Parijojana (MKSP) is committed to fulfil the specific needs of women farmers.

3.6.3 *Jeevika*

Jeevaka, a joint programme of World Bank and the Bihar Government, was launched in 2006 to improve the social and economic status of Bihar's poor. The modus operandi of the scheme is to connect poor women living in rural areas to Self-Help Group (SGH) in a way to create self-employment and to reduce the dependence on exploitative private money lenders. The project targets poor household's women with one female representative from each household, 20–30 SHGs were formed in 1 village. Further, the project promotes the participation of women in producer organization. Women get engaged in various economic activities such as production of incense stick, honey production, and in allied sector of agriculture.

SHGs along with producer organizations create a platform for community investment and facilitate credit through formal institutions. It also provides technical assistance to the society at scale. Consequently, women get rid of the exploitative debt trap and moved to the formal sector and make their livelihood better.

In 2006, Jeevika was implemented in five districts (Muzaffarpur, Nalanda, Madhubani); in 2007, the project added 18 new blocks of six districts (Madhubani, Muzaffarpur, Gaya, Nalanda, and Khagria), and other 24 blocks of mentioned districts were added in 2009. In 2010, Bihar government converged the Jeevika programme with Government of India's scheme DAY-NRLM and notified Jeevika as Bihar State Rural Livelihood Mission. Till the target of MDG 2015, the programme had identified 3 lakhs community professionals and 75 thousand resource persons, mobilized 1.5 crore rural households into 10 lakh SHGs. The programme empowered rural women of Bihar by creating platforms for village organization (65,000), cluster-level federation (1600) and block-level federation (534).

3.6.4 MDG2

The Universal Elementary Education (UEE) plays an important role in strengthening the social fabric of democracy through provision of equal opportunities to all, which has been accepted since the inception of our Republic. Bihar is ahead of its target to ensure that by 2015 children everywhere, boys and girls alike, will be able to complete a full course of primary schooling as reflected in Fig. 3.4, Bihar has crossed the mark of 100% in terms of the total GER in 2007, and for boys it was done earlier in the year 2005. Mukul (2015) in his report said the major achievement of Bihar Government was the Net Enrolment Ratio that reached 93.77% at the primary level, 87.63% at upper primary level (from last year's 79.06%), 42.08% at secondary level, and 19.06% at higher secondary level. Bihar has made remarkable progress in bringing 23 lakh out-of-school children of 6–14 age group into schools. In 2005, approximately 12% of children were out of school, which has now come down to 2% (Government of Bihar 2017).

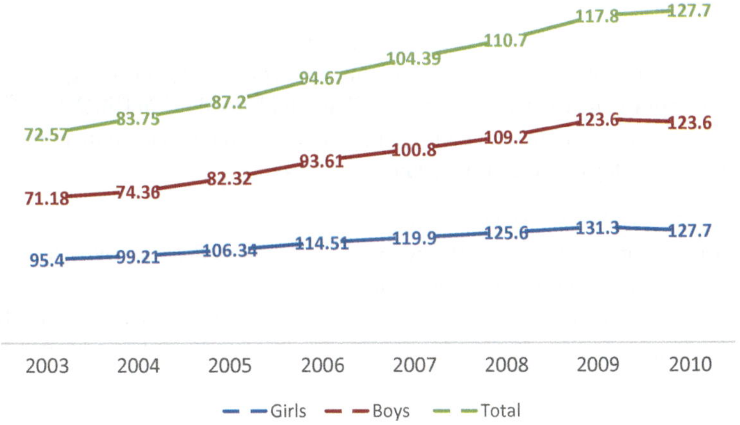

Fig. 3.4 Gross enrolment ratio under Primary School, Bihar. *Source* NITI Ayog (2015)

NER is calculated as the number of children of official primary school age, who are enrolled expressed as a percentage of the total population of children of official primary school age. GER is calculated as total enrolment in primary education, regardless of age, expressed as a percentage of the eligible official primary school-age population in an academic year, as reflected in Fig. 3.4, more than 100% values show that the country/state successfully accommodated all its school-age children.

3.6.5 Education as Fundamental Right

Education comes under the concurrent list of the Indian Constitution, making Center and States as equal partners in framing educational policies. However, union exercises supremacy over the states in enacting education-related laws; with the executive using its power to issue directions to States. The States' powers are contained to the extent that it does not stall the exercise of executive powers of the Union. Withstanding the pace of progress achieved till 2000, many programs have been formulated and implemented since 2001 to advance the goal of education for all. Such policies have been implemented by the joint efforts of Center and States by means of district-level decentralized management structures and local bodies. **Charwaha Vidyalayas** (school for poor cattle grazers).

In rural India, children in poor families are burdened with multiple tasks, from caring of younger siblings to taking animals for grazing. To be benefited in kind, they also escort the animals of others. Taking these circumstances, the *chrwaha* vidvalay was established in Bihar. The "*charwaha vidyalayas*" was a brainchild of the prominent socialist leader and champion of social justice, Sh. Lalu Prasad Yadav. The first school was inaugurated in 1993, during his early years in power for those children who tended to animals and could not attend regular schools.

This "charvaha vidayalya" gives a unique education system; it provides education to children from families with scanty income. It incorporated the system of "earn while you learn" by allowing the children to tend to the flocks that they brought out for grazing while attending classes. This was the pre-right to education era, where boys and girls in crumpled white–brown uniforms with goats, buffaloes, and cows in tow, enter a sprawling roadside campus at Charvaha Vidyalaya in Bihar. Setting their cattle loose, they assemble under a huge banyan tree and chant an anthem penned by their Chief Minister Laloo Prasad Yadav.

Ay suar charanewalo/Ay bakri charanewalo
Ay bhains charanewalo! Ay gai charanewalo
Ay ghonga chunnewalo/Ay kagaz clumnewalo
Ay shisha chunnewalo,
Padhna likhna sikho
Padhna likhna sikho

(All pig grazers/goat grazers, buffalo grazers/cow grazers, snail pickers/paper pickers, glass pickers, learn to read and write).

The school provide an opportunity for the children of vulnerable and marginalized community, who are engaged in domestic chores or caring animal herds. The campus is unique in that they provide grazing field to the cattle accompanying the students. Students are also provided with mid-day meal.

The student is approached by block officials and lured to join the Charwaha Vidyalaya by the dazzling prospect of a free midday meal, two uniforms, books, and a monthly stipend of Rs 9. And the clincher: a lush green pasture on which to graze his cattle, with fodder to carry home at the end of the day. More remarkable, perhaps, is that nearly half the students (46%) are girls. All are children who, in the normal scheme of things, would never have tasted that sadly derisory birthright"-a school education. The "musher" community constitutes 2.19% of Bihar population was the main target group of this scheme. Despite being a visionary idea, the Charwaha schools failed due to the lack of practical long-term planning. There was no separate budgetary allocation to run them.

3.6.6 Right to Education Act 2009

To universalize elementary education, the Government of India implemented the Sarva Siksha Abhiyan (SSA) along with RTE ACT 2009 in 2010. Education leverages progress and development. Providing free education is a vital state responsibility (Sen 2003). The Government of India implemented a historic law called Right to Education (RTE) to provide free and compulsory education to all children in the age group of 6–14 years. With the 86th Constitutional amendment, education became a fundamental right in 2002 (Government of India 2002). The Right of Children to Free and Compulsory Education Act, a law to enable the implementation of the fundamental right, was passed by Parliament in 2009. Both the Constitutional amendment and the new law came into force from 1st April 2010 (ToI, April 2010). RTE act is not only about universal enrolment of children in schools, but also includes the improvement of teaching quality, school infrastructure teacher–student ratio, and infrastructural development required for betterment of education in India.

Taking cognizance of Section 38 of RTE Act 2009, the Bihar government introduced the "Bihar State Free and Compulsory Education of Children rule-2011" (Government of Bihar 2011). The state is legally bound by RTE to ensure free and compulsory education to every child. The Act discourages the practice of admission denial on the basis of caste, creed, and religion; while providing that only trained teachers are appointed by the schools. RTE Act emphasizes on school infrastructure such as availability of separate toilets for boys and girls, classrooms, facility of drinking water, pupil–teacher ratio, and working hours for teachers and made it mandatory for all the elementary school in India. The Act ensures continuing education with provision that no pupil is failed or expelled till 8th standard.

3.6.7 Performance of Goal

A person aged 7 years and above who can both read and write with understanding in any language has been taken as literate in India. It is not necessary for a person to have received any formal education or passed any minimum educational standard to be treated as literate.

The literacy rate for the population in the age group of 15–24 years has shown an upward trend both at the national and state levels, for females as well as males. In Bihar, the youth literacy rate has increased from 57% in 2001 to 72.3%. In the case of female education, it takes a big jump of 20.7 and 10.6% change is seen in male literacy rate (Census 2011). The bright line is that from 2001 to 2011 the increase in youth literacy rate was higher among females than males and the gender gap is also declining.

At primary level, total enrolment in primary increased from 146,292,333 (2007–08) to 16,170,088 children in 2015–16. Improvement in student's classroom ratio was also registered to 51 from 96 during the same period (UNICEF 2016) (see Figs. 3.5 and 3.6).

The cases of dropouts also decreased during 2006–07 to 2012–13 registering 14.4% reduction at all levels of education and 15.1% at primary level. The improvement in education sector shows the positive impact of SSA and RTE Acts in universalizing education and betterment of intrastate.

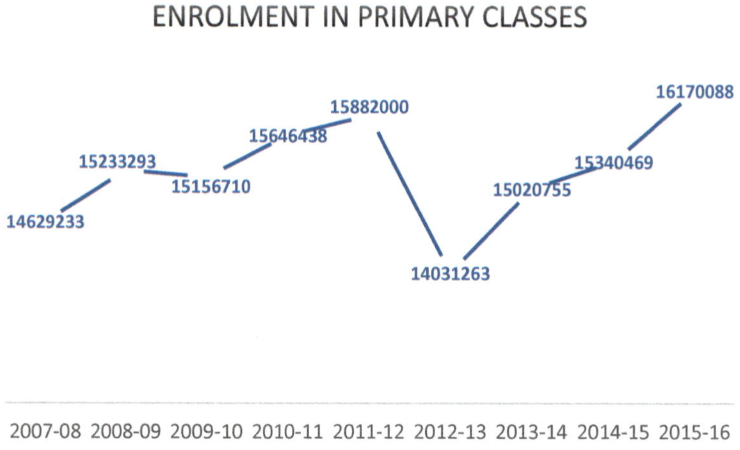

Fig. 3.5 Enrolment in primary classes (Bihar). *Source* UNICEF (2016)

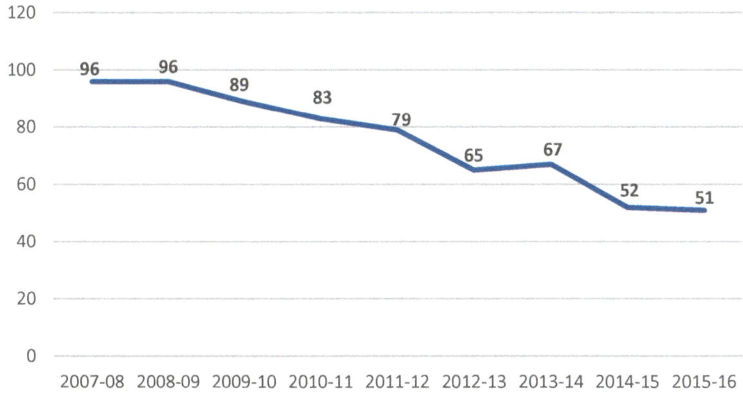

Fig. 3.6 Students–classroom ratio (Bihar). *Source* UNICEF (2016)

3.6.8 MDG3

Gender parity is of seminal importance for human development (Sen 2001). Without the full participation of half the population, the realization of any developmental goal is impossible. Gender equality is considered a human right. Women empowerment is identified as a prerequisite for poverty reduction. An empowered woman invests more in their children's education, nutrient food, which consequently facilitate the growth of family. Although evidence support the positive role of women in achieving human development, still gender parity is an unrealized dream (UNFPA 2016).

Policy-makers at the central and state levels are working tirelessly to ensure gender equality. The efforts of the governments at both levels are visible in the literacy trend of Bihar. The women literacy in Bihar increased to 51.5% in 2011 from 33.1% in 2001 (see Table 3.8). The decadal change was impressive being 18.4% for women while the change for men registered only 11.5% (Census 2001, 2011).

Gender disparity begins right at birth. Female infants are considered as a burden on the family. They receive less care during the initial year of birth and thus increasing the chances of infant mortality. The patriarchal society of Bihar limits the education opportunities for girls, which is highly unequal, social and economic background being hurdles in attainment of education. The first indicator of MDG3 is dedicated to Gender Parity Index (GPI) and the proportion of girls to boys in primary, secondary, and tertiary levels of education is measured by Gross Enrolment Ratio (GER), number of students enrolled in certain level of education, irrespective of age, presented as the proportion of official school age population corresponding to the same level of education. Data shows that the GPI for Bihar lagged behind and fell short of the target. (AISHE 2019) (see Fig. 3.7).

The second indicator measure the proportion of literate women to men under the age group of 15–24 years. The United Nations has identified the same age group as the youth population, unbiased to any other definitions given by member state. The

Table 3.8 Literacy rate, Bihar

Districts	2001			2011			Decadal change		
	Person	Male	Female	Person	Male	Female	Person	Male	Female
W. Champaran	38.9	51.1	23.2	55.7	65.6	44.7	16.8	14.5	19.5
E. Champaran	37.5	49.3	24.3	55.8	65.3	45.1	18.3	16.0	20.8
Sheohar	35.3	45.3	23.9	53.8	61.3	45.3	18.5	16.0	21.4
Sitamarhi	38.5	49.4	26.1	52.1	60.6	42.4	13.6	11.2	16.3
Madhubani	42.0	56.8	26.3	58.6	70.1	46.2	16.6	13.3	19.9
Supaul	37.3	52.5	20.8	57.7	69.6	44.8	20.4	17.1	24.0
Araria	35.0	46.4	22.4	55.5	63.7	46.8	20.5	17.3	24.4
Kishanganj	31.1	42.7	18.6	55.5	62.3	43.9	24.4	19.6	25.3
Purnia	35.1	45.6	23.4	51.1	59.1	42.3	16.0	13.5	18.9
Katihar	35.1	45.3	23.8	52.2	59.4	44.4	17.1	14.1	20.6
Madhepura	36.1	48.8	22.1	52.3	61.8	41.7	16.2	13.0	19.6
Sahrasa	31.9	51.7	25.3	53.2	63.6	41.7	14.1	11.9	16.4
Darbhanga	44.3	56.7	30.8	56.6	66.8	45.2	12.3	10.1	14.4
Muzzafarpur	48	59.1	35.8	63.4	71.3	54.7	15.4	12.1	18.9
Gopalganj	47.5	63	32.2	65.5	76.5	54.8	18.0	13.5	22.6
Siwan	51.6	67.3	36.9	69.5	80.2	58.7	17.9	12.9	21.8
Saran	51.8	67.3	35.8	66	77.0	54.4	14.2	9.7	18.6
Vaishali	50.5	63.3	36.6	66.6	75.4	56.7	16.1	12.1	20.1
Samstipur	38.5	49.4	26.1	52.1	60.6	42.4	13.6	11.2	16.3
Begusarai	48	59.1	35.6	63.9	71.6	55.2	15.9	12.5	19.6
Khagaria	41.3	51.8	29.4	57.9	65.2	49.6	16.6	13.4	20.2
Bhagalpur	49.5	59.2	38.1	63.1	70.3	54.9	13.6	11.1	16.8
Banka	42.70	55.3	28.7	58.2	67.6	47.7	15.5	12.3	19.0
Munger	59.5	69.9	47.4	70.5	77.7	62.1	11.0	7.8	14.7
Lakhisarai	48	60.7	34	62.4	71.3	52.6	14.4	10.6	18.6
Sheikhpura	48.6	61.9	33.9	63.9	73.6	53.4	15.3	11.7	19.5
Nalanda	53.2	66.4	38.6	64.4	74.9	53.1	11.12	8.5	14.5
Patna	62.9	73.3	50.8	70.70	78.5	62.0	7.8	5.2	11.2
Bhojpur	59	74.3	41.8	70.5	81.7	58.0	11.5	7.4	16.2
Buxar	56.8	71.9	39.9	70.1	80.7	58.6	13.3	8.8	18.7
Kaimur	55.1	69.7	38.8	69.3	79.4	58.4	14.2	9.7	19.6
Rohtas	61.3	75.3	45.7	73.3	82.9	63.0	12.1	7.6	17.3
Aurangabad	57	1.1	41.9	70.0	80.1	59.7	13.3	9.0	17.8
Gaya	50.4	63.3	38.7	63.7	73.3	53.3	13.3	10.0	16.6
Nawada	46.8	60.6	32.2	59.8	70.0	48.9	13.0	9.4	16.7

(continued)

Table 3.8 (continued)

Districts	2001			2011			Decadal change		
	Person	Male	Female	Person	Male	Female	Person	Male	Female
Jamui	42.4	57.1	26.3	59.8	71.2	47.3	17.4	14.1	21.0
Jehanabad	55.3	70.1	39.4	66.8	77.7	55.0	11.5	7.6	15.6
Arwal	NA	NA	NA	67.4	79.1	54.9	NA	NA	NA
Total	47.0	59.7	33.1	61.8	71.2	51.5	14.8	11.5	18.4

Source Census (2001, 2011)

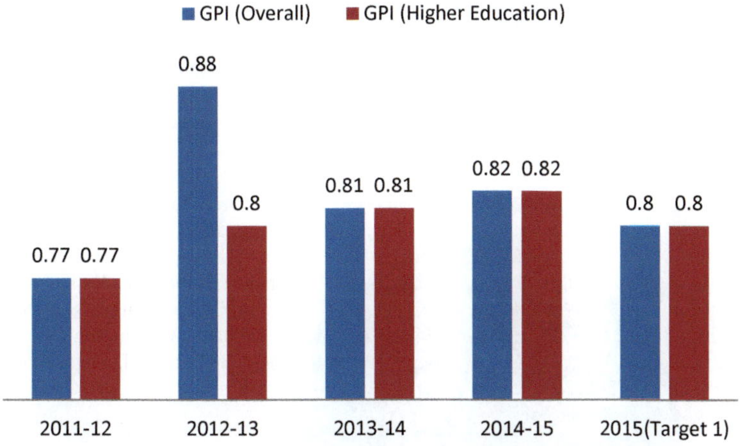

Fig. 3.7 Gender parity index, primary and tertiary level. *Source* AISHE (2019)

proportion of youth literacy, female to male in Bihar increased to 17% in 2011 from 13.75 in 2001.

The third indicator deals with the share of women in nonfarm wage employment. Like all the developing economies, most of the women in Bihar are employed in farm activity. Although, the number of women in nonfarm sector is very low in Bihar. Societal tendency and traditional values restrict the women participation in non-farm work.

Occupancy of women in national parliament, state legislative, and local levels of governance is the fourth indicator. In the general election of 2014, representation of women in Lower House (Lok Sabha) is reached to 68/573, i.e., 11.87% of total occupancy, in the same house Bihar had sent 3/41 women Parliamentarian, i.e., 7.32%, below the national average (see Fig. 3.8) (Lok Sabha 2016).

"Sarva Siksha Abhiyan" (SSA), provides umbrella cover to all the programs related to the upliftment of education to improve the enrolment ratio of girls over boys. Bihar Education Project (BEP) Council and SSA together benefited 2.6 crore children in the age group of 6–14 years. BEP along with SSA work for other programs, like National Program for Education of Girls at Elementary Level (NPEGEL), a

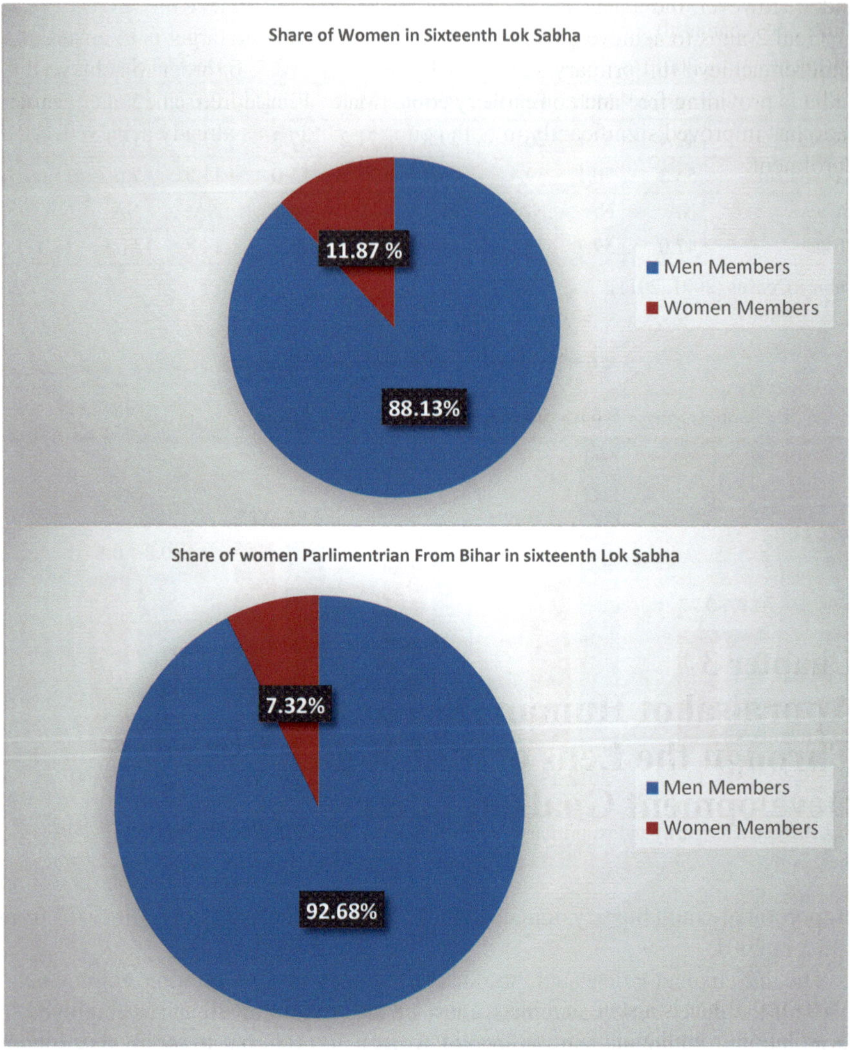

Fig. 3.8 Women representation in Sixteenth Lok Sabha and share of Bihar. *Source* Lok Sabha (2016)

special component of SSA which provides additional support for enhancing girls' education, over and above the normal SSA intervention.

3.6.9 National Program for Education of Girls at Elementary Level (NPEGEL)

The National Program for Education of Girls at Elementary Level (NPEGEL) provides for the setting up of model schools in every nook and cluster to address the issue of girls, who are irregularly in and out of school periodically. The goal is to increase girls' enrolment in schools by means of community supervision, awareness, and mobilization. This scheme is implemented in Educationally Backward Blocks (EBBs), where rural female literacy stands below the national average of 46.1%; with a gender gap above the national average of 21.6% (Census 2011). The EBB Scheme has been transferred to the state government from the center from 2015 to 2016.

The first few years after birth is extremely important for the child because at this delicate age children are vulnerable to growth retardation, micronutrient deficiencies, and common childhood illness. It is the phase when one million neural connections are generated each fraction of second, the frequency never repeats again (UNICEF 2017). Undernutrition in early stage invites diseases and causes child mortality. IIPS (2021) report assessed the weight of children under five and found that 43% of children experienced stunting and uneven growth due to undernutrition. An estimated 23% are wasted and 9% are cruelly wasted because of inadequate food intake. Almost 41% of children are malnourished, signaling severe undernutrition.

Bihar has shown progress in child care during MDG period. The child stunting improved to 43% in NHFS-5 from 47% in NFHS-4. The share of underweight children shrinks to 41percent from 44% during the same period, as shown in Fig. 3.9.

As compared to national level, as per the latest available data (Fig. 3.10), the proportion of underweight among children < 3 years of age is alarming in Bihar with 54.9%, while the proportion of underweight children < 3 years is more than the national level estimate of 40% in 2005–06. Bihar set the target to bring the percentage of underweight children below 3 years down to 27% in 2015 (NFHS-2, 1998–99 and NFHS-3, 2005–06). Bihar has shown improvement in reduction of proportion of under-weight children below 3 years of age but has failed to achieve the target with the proportion recorded at 43.3% against the target of 26%.

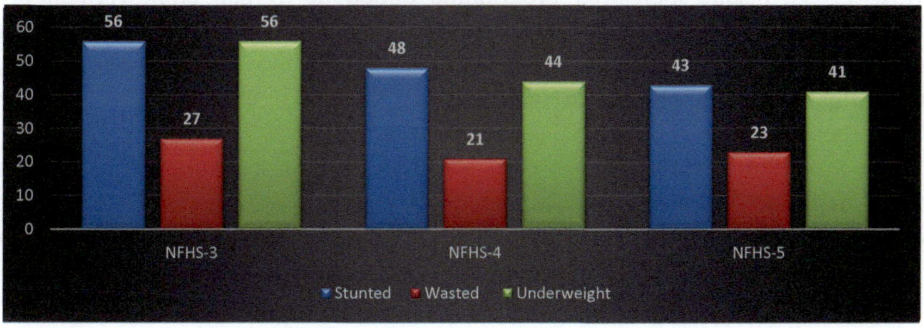

Fig. 3.9 Trends of malnutrition (Bihar). *Source* NFHS-3, NHHS-4, and NHHS-5

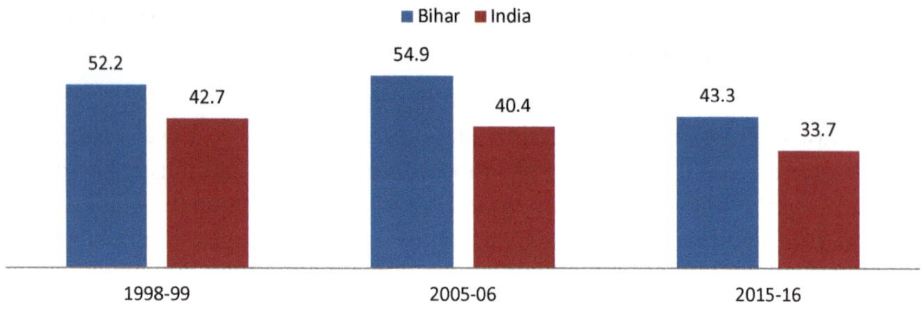

Fig. 3.10 Proportion of under-weight children below 3 years of age. *Source* Country Report (2017)

3.6.9.1 Integrated Child Development System (ICDS)

With more than 35 years of implementation, ICDS is one of the world's largest and unique program for early childhood care and development. It shows the government's commitment to its children and nursing mother. ICDS programme provides supplementary nutrition, monitoring growth, health education, vaccination, initial school education, and nutrition consultation. ICDS performs these services through huge network of *anganwadis*, managed by workers in assistance of helpers. ICDS was conceived as a response to the challenges of providing preschool and nonformal education on one hand and breaking the vicious cycle of malnutrition, morbidity, reduced learning capacity, and mortality on the other. The children aged between 0 and 6, pregnant women, and lactating mothers are the beneficiaries of this universal scheme. Each *anganwadi* is tasked to cover 200 families, almost a population of 1000 . ICDS services are designed to address the multiple challenges of child care (Gragnolati et al. 2006). Adhikari and Bredenkamp (2009) studies shows that the investment ICDS programme is continuously increasing over the year.

3.6.9.2 Mukhya Mantri Nari Sakti Yajana (MMNSY)

MMNSY is a state-sponsored programme aimed at women empowerment by strengthening women socially, culturally, and economically. The programme has special provision for victims of domestic violence by providing protected shelters, short stay houses and palnaghars of children. It also promotes inter-caste marriage in the society.

The programme covers 6131 villages, 1532 Panchayats constituted in 165 blocks of 27 districts. It works with women development corporation to connect the women of marginalized and vulnerable communities to the main stream economy by mobilizing them on platform provided by SHGs for their overall development.

3.6.9.3 Kasturba Gandhi Balika Vidyalaya (KGBV)

This programme named after Mahatma Gandhi's wife Kasturba Gandhi aims to promote education among the weaker section of the society by establishing residential schools for the girls belonging to SC/ ST, OBC, and minority communities. Under KGBV, Government of Bihar has constructed 533 schools out of which 46,000 girls are enrolled in 530 schools (Government of Bihar 2015).

3.6.9.4 Mukhyamantri Balika Cycle Yojana (MBCY)

MBCY is a bicycle programme for girls. The programme has transformed women education in the state. The scheme provides Rs. 2500 to every female candidate who passed 8th standard to purchase a bicycle so that they can commute regularly to school. The scheme helped in reducing the dropout rates among the girl students.

3.6.9.5 Mukhyamantri Balika Poshak Yojana

Under the "*Mukhyamantri Balika Poshak Yojana*", all girls studying in classes 6 to 8 are given Rs. 700 each for school uniforms and study material. "Over 3.6 million students have benefited from the scheme since it began in 2006". In 2009, the scheme was extended to girl students of classes 3, 4, and 5. 57 lakh girls studying in classes IX to XII who have been provided monetary help for purchasing school dress under the "*Mukhtamantri Balika Poshak*" Yojna since the year 2011–12 till 2015 (DoE: Bihar 2015).

3.6.9.6 Bihar Bal Bhawan "Kilkari"—Bihar Children House

This scheme is available for children of disadvantage groups cannot afford the facility of proper schooling. In 2008, the Bal Bhawan "Kilkari" was established to improve elementary education and encourage students to learn. This institution has been awarded at national level for its work.

3.6.9.7 "Gyan Jyoti" Program

Under the "Gyan Jyoti" program about 72,000 girls were imparted training in local vocational training and theater skill training was imparted to about 76,000 girls (Government of Bihar 2015).

3.6.9.8 Mukhyamantri Kishori Swasthya Karyakram

"Mukhyamantri Kishori Swasthya Karyakram" covers girls from Class VII to Class XII and nearly four million schoolgirls in government-run middle and high schools in Bihar were provided free sanitary napkins since the inception of this program in 2010.

3.6.9.9 Mukhyamantri Kanya Vivah Yojana

Mukhyamantri Kanya Vivah Yojana is operational since 2007–08. The scheme provides financial assistance to poor families at the time of marriage of girls who have attained the age of 18. The programme facilitates marriage registration, prevention of child marriage, and promote women education. As per Bihar's Government report (2015), between 2007 and 08, 78,385 beneficiaries were registered.

3.6.9.10 Mukhya Mantri Kanya Suraksha Yojana

Under the Mukhya Mantri Kanya Suraksha Yojana, the state government donates Rs 2000 each on the birth of the first two baby girls in BPL families (Business Standard 2013). The program's objective has been to ensure the safety and security of the girl child, instill pride for the girl child in the minds of the society, improve the overall sex ratio, and promote registration of birth. The program is said to have brought out a change in the mindset of the poor regarding. The state government has rewarded couples celebrating the birth of girl child in the family under its Prerna Award Scheme.

3.6.9.11 The Bihar Panchayat Raj Act, 2006

The Bihar government has amended its own Panchayat Raj Act 1993 to make fifty percent women reservation at third tier of democracy. Reservation of seats are done in the manner that, in every Gram Panchayat, as nearly as but not exceeding fifty percent of the total seats of members of Gram Panchayat shall be reserved for women. Clause 13 (20) of the Act made it clear that this reservation shall be applicable with the reservation pattern of other social groups (The Bihar Panchayat Raj act 2006).

3.6.9.12 MDG4: Reduce Child Mortality

The first indicator under MDG4 is to check the under-five mortality rate (U5MR). In India, the Office of the Registrar General of India (ORGI) regularly releases reliable estimates of fertility and mortality using data collected through Sample Registration System (SRS). As presented in Fig. 3.11, under-five mortality ratio (U5MR) in Bihar

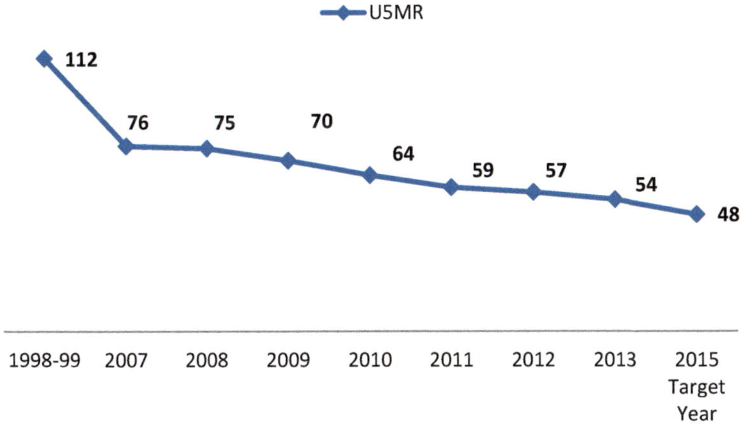

Fig. 3.11 Trends of Under-five mortality rate, Bihar. *Source* SRS Report (2013), MDG Final Report (2017)

was estimated at 75 deaths per 1000 live births in 2007 (SRS, 2007–13). In order to achieve Target 5, the U5MR is to be reduced to 42 deaths per 1000 live births by 2015. As per the India's MDG final report, U5MR is at 48 deaths per 1000 live births. However, an overall reduction in U5MR happened during 2000–2013, registering a faster decline in the recent past, and if this rate of reduction is sustained the achievement by 2015 is very close to the target by 2015 (MoSPI 2017c).

On the basis of the Bihar Annual Health Survey-2012–13 as conducted in 37 districts, the state can take pride in the fact that it registered the sharpest fall in infant mortality rate (IMR) among the nine surveyed states. It stands at 48, as per latest AHS bulletin, as against 52 in 2011–12. As many as 16 districts recorded IMR above state value while another 16 recorded below it. Patna (31) has the lowest IMR while Madhepura (64) has the highest followed by Sitamarhi, Supaul, Kishanganj, and Khagaria. Female infants die more than males (as depicted in Fig. 3.12). The high neonatal (less than 29 days of birth) mortality still continues to be a significant contributor to the infant mortality rate in Bihar. SRS reports (2013) that neonatal

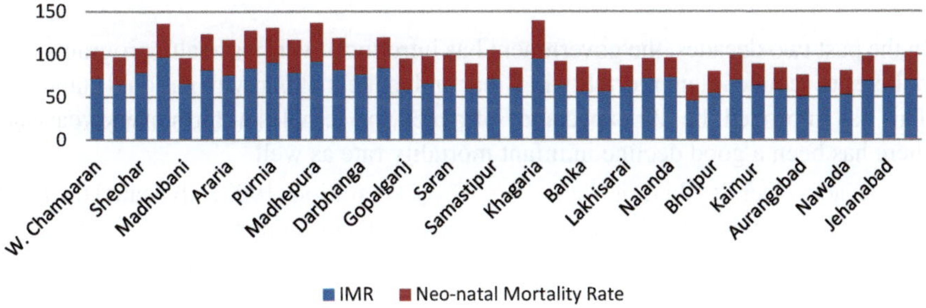

Fig. 3.12 Districts trends of infant mortality and neonatal mortality, Bihar. *Source* AHS (2013)

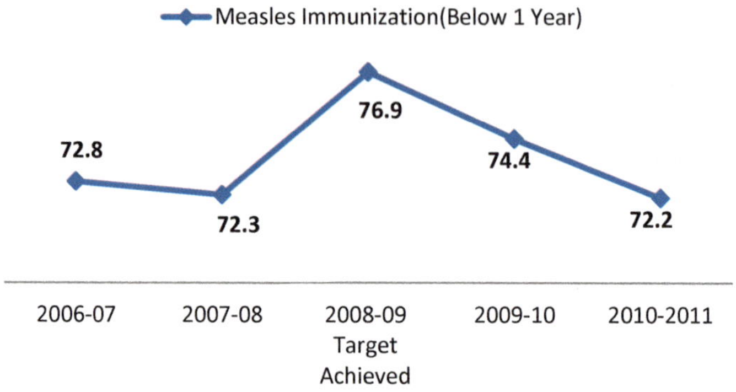

Fig. 3.13 Immunization against Measles, Bihar. *Source* Ministry of Health, GoB

mortality in Bihar ranged between 18 in Patna and 45 in Madhapur, the overall rate of Bihar is 32. Bihar has failed in attaining the target of 27 infant deaths per 1000 live births by 2015.

Fourth indicator is to bring the proportion of children below one year immunized against measles. In India, maximum child mortality is recorded in the age group of 0–5, and most of the diseases are prevented by immunization. Trends shows that Bihar has expanded the cover of universal immunization in universal immunization (see Table 3.8). State has achieved complete inoculation for TT (139%). With current trends state is on path to achieve the immunization targets for BCG, OVP, and measles in near future. Although, in 2015, Bihar proudly declares itself as tetanus-free state for maternal and neonatal.

The pentavalent vaccination in which five individual vaccines has been conjugated (homophiles influenza type B, whooping cough, tetanus, hepatitis B, and diphtheria) has started in 2015 which safeguards the infant from those five deadly diseases. Going by their historical rate, as presented in Fig. 3.9, of increase in coverage, Bihar has achieved the universal coverage in measles immunization of one-year-old children by 2015 along with India (Annual Health Survey 2013) (Table 3.9).

3.6.9.13 Program to Keep Check on Child Mortality

In the past two decades, the government has introduced various health programs and policies to reduce infant mortality in Bihar and make it at par with national standard. This has improved the life expectancy of males and females at birth every year and there has been a good decline in infant mortality rate as well.

Table 3.9 Trends of universal coverage in measles immunization of one-year-old children

Antigen	2012–13			2013–14			2014–15		
	Goal	Success (000)	Success (%)	Goal	Success (000)	Success (%)	Goal	Success (000)	Success (%)
TT (P.W.)	3121	4460	143	3106	4306	139	3184	2273	71
BCG	2916	1597	89	2926	2685	91	2999	1208	40
Measles	2916	2244	77	2926	2330	80	2999	1331	44
OPV1	2916	1990	68	2926	1801	62	2999	834	28
OPV2	2916	2601	89	2926	2483	85	2999	1093	36
OPV3	2916	2499	86	2926	2461	84	2999	1141	38
DPT1	2916	2384	82	2926	2400	82	2999	1202	40
Penta1		2603	89	2926	2684	92	2999	1101	37
DPT2 Penta 2	22,916	2503	86	2926	2659	91	2999	1167	39
DPT3 PENTa3	32,916	2388	82	2926	2580	88	2999	1241	41

Source Government of Bihar (2016)

3.6.9.14 Muskaan Ek Abhiyan (The Smile Campaign) for Strengthening Routine Immunization in Bihar

The state has initiated a new program called "MUSKAAN EK ABHIYAN", which was launched in October 2007 under the National Rural Health Mission (NRHM) to give a fillip to the immunization program in Bihar by tracking pregnant women and new born child. Under this program, ASHA, AWW, and ANMs jointly track the pregnant mothers and new born child. Under this program, ASHA, AWW, and ANM hold meeting with "Mahila Mandals" in AWCs. The main objective is to cover ANC coverage and immunization. As reported, after the initiation of this program, the coverage of ANC and Immunization has been enhanced. The ANM and ASHA are being given incentives to promote this program.

3.6.9.15 Supplementary Nutrition Program (SNP)

The State Government of Bihar allots a fund of USD 200M every year to combat child malnutrition through its Supplementary Nutrition Program (SNP) run by the Integrated Child Development Services (ICDS) Directorate of the Department of Social Welfare. The SNP program provides funds directly to village-level anganwadi (pre-school) workers to buy, cook, and serve food to children in the age group of 6 months to 3 years; as well as pregnant and lactating mothers. However, the government reports alleges 50% spending being diverted; meals provided only 60% of the days with missing nutritional content as stipulated originally by the government (GoB Report Card 2015).

3.6.9.16 MDG5: Improve Maternal Health (MMR)

The MDG5 stipulates that the Maternal Mortality Ratio (MMR) level be reduced by three-fourths between 1990 and 2015. In 1990, the estimated MMR was 437 per 1,00,000 live births. In order to meet the MDG target, the MMR should be reduced to 109 per 1,00,000 live births by 2015.

The first indicator is maternal mortality ratio and as per the SRS report during 2004–13, 33.3% change is visible in the MMR of Bihar (see Fig. 3.14). The MMR of Bihar in the period of 2004–06 is 312, which came to 208 during 2011–2013. During the period 2004–13, MMR in Bihar declined by 104 points, whereas at all India level the decline was 87 points (Office of Registrar General of India 2015). Thus, the states with highest level of MMR in 2011–13 have shown maximum points of decline, which offers a ray of hope in the progress of states toward achieving MDG5.

Second indicator of MDG5 is the "Proportion of births attended by skilled health Personnel". One of the important dimensions of health infrastructure is the availability of skilled health personnel [doctors, nurses, Auxiliary Nurse-cum-Midwife (ANM), an Accredited Social Health Activist (ASHA)]. The district-wise number

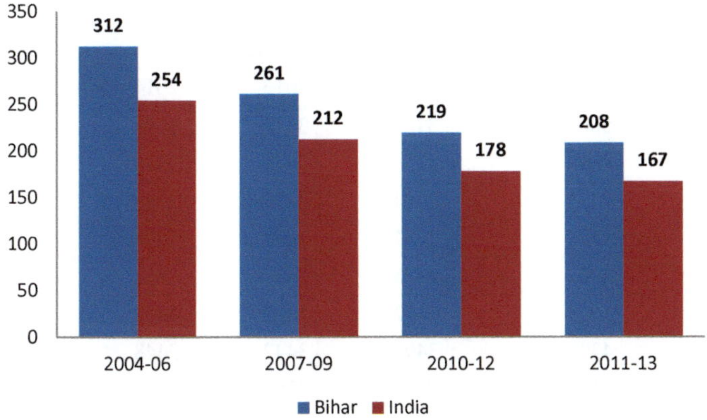

Fig. 3.14 Evidences of maternal mortality rate India and Bihar. *Source* Country Report (2017)

of ASHA workers (sanctioned and at present at work) is presented in Table 3.10. "The state government has 85,045 ASHA workers in 2014–15 against the 4447 in 2012–13. The total number of allotted post (contractual) is 10,423" (State Health Society: GoB 2015–16).

An evaluation study endorsed by Planning Commission has pointed out that more than 65% of beneficiaries are visited by ASHA workers once in 15–30 days while 80% of beneficiaries have confirmed delivery of free drugs by ASHA. On the other hand, around 65% of beneficiaries have opted for institutional delivery and 60% of women have availed antenatal care services in government facilities highlighting the crucial role played by ASHA workers in encouraging pregnant women to utilize public healthcare facilities. 56% of couples reported using contraception of which 88% was availed from government health counters (PIB 2015). ASHA has also been helpful in educating patients to avail facilities of public health institutions for chronic diseases than private ones. The evaluation study also reflects the need to increase payment compensation to ASHA's and provide incentives for emergency support.

The new initiatives create a database of ASHA workers in all states, further facilitating the system for outcome monitoring. A handbook for ASHA facilitators has been introduced and ASHA grievance redressal mechanism has been set up in Bihar and other states which shall surely come to their benefit.

Contrary to common assumptions, (from Tables 3.10 and 3.11), the strength of health personnel is much higher for ANM and ASHA. In 2015–16, the strength of regular ANM was 8895, against the sanctioned strength of 11,805, implying a vacancy ratio of 25%. Apart from regular ANMs, the state government has also sanctioned posts of 12,587 contractual ANM, and 9670 of them have already been filled. The ASHA figures amount to 850,000 of them working in the entire state, against a sanctioned strength of 1042 thousand, implying a vacancy ratio of only 18% (State Health Society, GOB 2015–16). When one compares the district-wise

Table 3.10 Distribution of ASHA workers in Bihar

District	Sanction post (contractual)	No of ASHA worker			ASHA/lakh population
		2012–13	2013–14	2014–15	
Patna	5842	3182	3018	3018	52
Nalanda	2880	2356	2313	2316	80
Bhojpur	2731	2264	2199	2188	80
Buxar	1708	1487	1474	1474	86
Rohtas	2962	2435	2430	2454	83
Kaimur	1628	1462	1462	1462	90
Gaya	4395	3493	3442	3448	79
Jehanabad	1126	871	870	870	77
Arwal	773	773	746	1956	107
Nawad	2221	1920	1928	2142	88
Aurangabad	2542	2158	2142	3402	84
Saran	3935	3422	3385	2822	86
Siwan	5099	2939	834	2374	85
Gopalganj	4805	2297	2336	3108	93
W. Champaran	3434	3064	3043	4060	79
E. Champaran	656	3762	3906	3858	80
Muzaffarpur	3498	3812	3856	2919	80
Sitamarhi	3941	2827	2912	572	85
Sheohar	4487	570	572	572	87
Vaishali	4266	2856	2888	3019	86
Darbhanga	3941	3546	3100	3118	79
Madhubani	4487	3632	3682	3852	86
Samastipur	4266	3829	3779	3794	89
Begusarai	2973	2361	2410	2403	81
Munger	1369	956	956	952	70
Sheikhpura	637	476	476	476	75
Lakhisarai	1002	756	756	901	90
Jamui	1762	1513	1504	1505	85
Khagaria	1668	1409	1399	1468	88
Bhagalpur	3040	2236	2236	2232	73
Banka	2037	1814	1782	1809	89
Saharsa	1902	1472	1471	1471	77
Supaul	2229	1920	1912	2093	94
Madhepura	2004	1638	1655	1674	84
Purenia	3268	2696	2634	2709	83

(continued)

Table 3.10 (continued)

District	Sanction post (contractual)	No of ASHA worker			ASHA/lakh population
		2012–13	2013–14	2014–15	
Kishanganj	1690	1334	1280	1290	76
Araria	2812	2362	2365	2365	84
Katihar	3074	2547	2549	2720	89
Bihar	**104,239**	**4447**	**83,702**	**85,045**	**82**

Source Health Department, GoB (2016)

Table 3.11 Number of health personnel, Bihar

Posts (name)	Regular				Contractual			
		Working			Post (sanctioned)	Working		
Doctors	6261	2392	2255	2052	2314	1580	1580	1488
Nurse (Grade A)	1688	397	356	1706	3502	1643	1621	412
ANM	11,805	9146	8999	8895	12,587	9484	9933	9670
ASHA	–	–	–	–	104,239	83,702	85,045	NA

Source Health Department, GoB (2016)

positions, variation is found to be limited regarding ANM and ASHA workers. But in terms of doctors and nurses, there is considerable inter-district variation.

Maternal health care is multidimensional and the services include at least three antenatal care visits, iron prophylaxis for pregnant and lactating women, at least one dose of tetanus toxoid vaccine, detection and treatment of anemia in mothers, and management and referral of high-risk pregnancies and natal care. Table 3.12 suggests the regular improvement in institutional deliveries in Bihar. These institutional deliveries are monitored regularly through Health Management Information System (HMIS) and also periodic surveys (State Health Society, GOB 2015–16).

3.6.9.17 Janani Suraksha Yojana (JSY)

Janani Suraksha Yojana (JSY) targets poor and marginalized women and aims to reduce the incidence of maternal mortality and neonatal mortality. The scheme provides for the reimbursement of the cost of deliveries in hospitals and also covers the transportation cost.

3.6.9.18 Indira Gandhi Matritva Sahyog Yojana (IGMSY)

IGMSY is a central government scheme dedicated to the health of lactating mothers. Any lactating mother above the age of 19 is eligible for the IGMSY scheme up

Table 3.12 Incidence of Institutional deliveries, Bihar

	2006–07	2007–08	2008–09	2009–10	2010–11	2011–12	2012–13	2013–14	2014–15
Delivery (institutional)	1.17	8.38	11.45	12.46	13.85	14.32	14.69	16.47	15.46
Arte of change (%age)	–	616.27	36.52	8.89	11.09	3.41	2.60	12.08	−6.10

Source Health Department, GoB (2016) (till September 2015) to the first two childbirths. The government understands that during pregnancy the women lose employment opportunities, to compensate for the loss and to keep the child and mother healthy, the government provides Rs 6000/- to enable them to feed themselves with nutritious food.

3.6.9.19 MDG6

The health status of a society is measured by the prevalence of major diseases. Most of the population suffered acute respiratory infection followed by fever of unknown origin, diarrhea, dog bite, dysentery, and enteric fever, the numbers are 17.3 lakhs, 13.2 lakhs, 6.9 lakhs, 6.3 lakhs, and 3.1 lakhs, respectively (Government of Bihar 2016) (see Table 3.13). The evidence shows that water-borne disease like diarrhea and dysentery are very common in Bihar. These diseases occur due to the lack of clean and safe drinking water.

At the national level, HIV prevalence among adults aged 15–49 years is estimated to have fallen from 0.34 to 0.26% from 2007 to 2015. Currently, adult prevalence is 0.30% among males, and 0.22% among females (NACO 2015). Bihar is among the list that has an adult prevalence of over 0.20%. The HIV estimation 2015 shows that India has been successful in achieving the MDG6 goal of halting and reversing the HIV epidemic by 2015. The fact that new infections have declined by 66% between 2000 and 2015, and AIDS-related deaths have fallen by 54% from 2007 to 2015 is concrete proof that India has delivered on its global commitment to achieving MDG6 (PIB 2015).

Authorities suggest that better text facility and awareness among the masses resulted in the rise of new cases. More turnout at pathology centers translated into

Table 3.13 Prevalence of major disease, Bihar

Diseases	2013–14	2014–15
Acute diarrhea	6.89 (11.7)	6.87 (13.1)
Bacillary dysentery	4.00 (6.8)	3.64 (6.9)
Viral hepatitis	0.14 (0.2)	0.26 (0.5)
Enteric fever	3.11 (5.3)	3.07 (5.9)
Malaria	0.33 (0.6)	0.29 (0.6)
Fever of unknown origin	16.43 (28)	13.189 (25.2)
Acute respiratory illnesses	21.52 (36.6)	17.30 (33.0)
Pneumonia	0.73 (1.2)	0.49 (0.9)
Dog bite	4.15 (7.1)	6.27 (12.0)
Any other state-specific disease	0.42 (0.7)	0.29 (0.6)
Unusual syndrome not captured above	1.03 (1.8)	0.71 (1.4)
Total	158.74 (100)	52.37 (100)

Source Government of Bihar 2016

a greater number of cases. The number of HIV cases increased to 64,000–65,000 in 2013 from 45,000 to 50,000 in 2012 (BSACA 2015), while the national agency NACO estimated 150,000 HIV cases in Bihar during the same period.

The second indicator is the proportion of the population who use condoms and contraceptive condoms (Condom use to overall contraceptive use among currently married women, 15–49 years). During NFHS-3 (2005–06), the total percentage of people using condoms as an interceptive was 2.3%, which changed to 3.1 during NFHS-4 (2015–16). In the same period, women who know that consistent condom use can reduce the chances of getting HIV/AIDS (percent) comes to 33.5% from 22.8 and 62.3–67% for men. (NFHS-4, 2015–16). These positive trends show the government's commitment to the completion of goals.

Bihar shows improvement in contracting malaria cases in the state. The state has recorded zero death due to malaria infection during 2011–2012. This progress is attributed to the National Vector Borne Diseases Control programme, which helped in achieving the target (NVBDCP 2015). Apart from the central government, World Bank assistance of $820.75 million facilitated the infrastructure for malaria detection and treatment in Bihar, which limits the expansion of infection in the state.

Tuberculosis, which is also resurgent worldwide, is an infectious disease that affects the lungs and other body tissues. Tuberculosis of the lamps, the most commonly known form, is characterized by coughing up mucus and sputum, fever, weight loss, and chest pain. According to WHO estimates, one-fourth of the global incidence of TB cases occurs in India annually with India accounting for 2.2 million, while the total incidence of TB is 9.6 million cases across the world annually (WHO 2015).

In Bihar, the incidence of TB cases is rising. NFHS-2 finding shows that 1% of the state population is infected with TB. Further, the report recorded a 0.6% rise in the case from the NFHS-1 survey to the NFHS-2 survey. The burden of diseases is more in rural areas in comparison to urban areas. The gap between rural and urban areas is 65% and the cases in rural areas recorded 1035/100,000 population while for urban areas the digit falls to 629/100,000 population (NFHS-4 2016).

In Bihar, the disease is more prevalent in males than females. Studies suggest that males are more prone to infection because men are working outside and are more likely to get infected by the people who carry the infection, another reason is smoking behavior in males is more than in women.

Fatality due to TB is very high, data shows that every 3-min TB takes 2 lives in India. As per the WHO report, TB incidence is falling in India. India has met the target of a 50% reduction in prevalence rate by 2015, compared with 1990 and a 50% reduction in mortality in mortality rate by 2015 compared with 1990 (WHO 2015).

3.6.9.20 MDG8: Environment and Sanitation

Integrate the principle of sustainable development into country policies and programs and reverse the loss of environmental resources.

3.6.9.21 Proportion of Land Area Covered by Forest

The state of Bihar presently has 6845 sq km of notified natural forest area, which is 7.27% of the geographical area of the state (see Fig. 3.15). Open forest constitutes 3.89%, moderate forest 3.59%, scrub percentage is 0.12 while the share of dense forest is merely 0.26%. These natural forests are spread in the districts of West Champaran, Kaimur, Rohtas, Aurangabad, Gaya, Jahanabad, Nawada, Nalanda, Munger, Banka, and Jamui. The north Bihar except West Champaran is devoid of natural forests. West Champaran is the only district with 231 km^2 of forests—which has a forest that is more than 70% dense (where it is almost impossible for sunlight to touch the ground). Sheikhpura (nil), Siwan (2 km^2), Buxar (3 km^2), and Gopalganj (4 km^2) are having the lowest share of forest in Bihar (Government of Bihar 2015).

Trends depicted in Fig. 3.16 suggest that the forest cover in Bihar is expanding by 0.47% between 2007 and 2013, while during the same period at the national level the change is only 0.18% (Environment, Forests and Climate Change 2013).

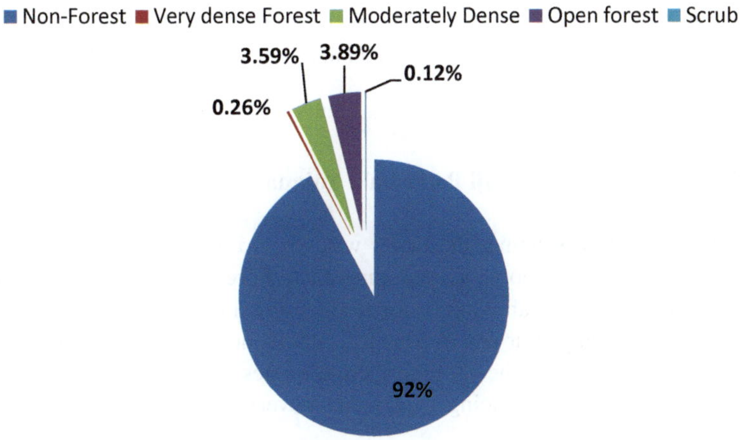

Fig. 3.15 Distribution of Forest area, Bihar. *Source* Country Report (2017)

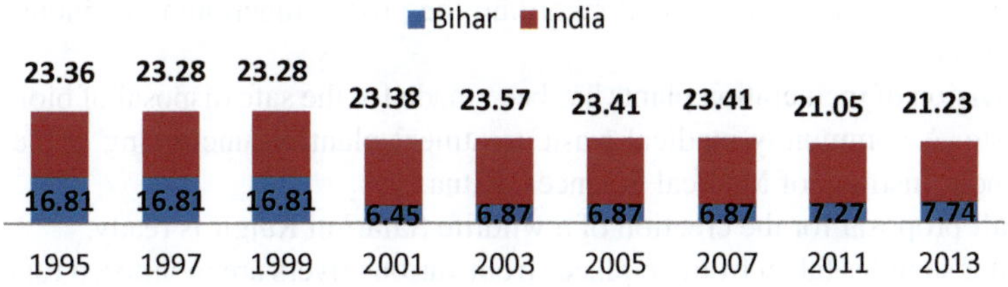

Fig. 3.16 Forestation trends of Bihar and India. *Source* Country Report (2017)

3.6.9.22 Swachh Bharat Mission-Gramin (SBGM)

Swachh Bharat Mission (Gramin) focuses on rural sanitation as it seeks to make the countryside healthier by promoting cleanliness, hygiene, and the achievement of open defecation-free status. This flagship programme of the government of India employs the Panchayati Raj institutions along with engaging the local communities to achieve the desired results. It also emphasizes on generating awareness regarding health education. Efficiency and effectiveness constitute the twin pillars of the programme as it relies heavily on the adoption of appropriate technologies that are ecologically sound and sustainable.

3.6.9.23 Har Parisar-Hara Parisar

Started in 2014, this is a scheme of town beautification through the plantation of trees. This scheme is an over-achiever as under the Scheme 1.71 lakh plantations have been planted across 234 campuses, as against the original target of 175 campuses. During the southwest monsoon of the year 2015–16, the implementing agency planted 604.2 lakh plants, achieving the target of 148.20 lakh plants by an astounding proportion (GoB Report Card 2015).

3.6.9.24 Mukhyamantri Niji Podhshala Yojana

Under this plantation scheme, the target was revised and raised to 507.66 lakhs, further benefitting 4314 people. (Environment and Forest Deptt, GoB 2015). Under the scheme, tissue culture laboratory cum production centers have been constructed at T.N.B. college in Bhagalpur and at Bharat Sevak College Supaul. Further, the state administration also established 8 high-tech nurseries in different forest divisions, each with a capacity of producing 1 lakh plants every year. The monitoring is done digitally.

- Under the Water (Pollution Prevention and Control) Act, 1974, and Air (Pollution Prevention and Control) Act 1981, 13,408 orders have been issued by the Bihar State Pollution Control Board regarding the proper functioning of industries in the state.
- Provision of incineration plants has been made for the safe disposal of biomedical waste. A community medical waste treatment plant is functioning at the Indira Gandhi Institute of Medical Sciences, Patna.
- Draft proposal for the creation of a wildlife Safari in Rajgir is ready.
- Water samples taken from 86 places from various rivers are monitored for quality assurance.
- Aranya Bhavan—a newly constructed multi-story building in Patna has been completed to ensure better functioning of the environment department.
- The government has established centers in Gaya, Muzzafarpur, and Patna for automatic air quality monitoring.

Access to safe and potable drinking water is an essential requirement for the sustenance of life on the planet. Water is the basis of all life on earth. Therefore, access to safe drinking water is justifiably recognized as a basic human right. Further, the argument for providing and safeguarding this essential human right becomes even stronger when the costs, both human and financial, associated with the lack of provision of safe drinking water are factored in. For example, according to 2014 WHO data, a total of 842,000 people die every year because of unsafe water or associated lack of sanitation, of which 361,000 constitute children under the age of 5. With respect to India, the World Bank data highlights that at the all-India level, 87.8% of households had an improved source of water in 2012. The rural and urban disparity is also highlighted.

Under the scheme of Indian federalism, providing safe drinking water to the citizens is the responsibility of the state government. With regard to Bihar, as per the last census (2011), the piped-water coverage in the state is a meager 2.5% of households. The government aims to take this number to 20% by 2022. For doing so, massive public investment will be needed.

Another key dimension related to sanitation is access to an improved source of latrine. As per the 2012 NSSO data, only 38.8 and 89.6% of households in rural and urban areas, respectively, have access to "improved source of latrine".

3.6.9.25 The National Rural Drinking Water Programme (NRDWP)

"Rural drinking water supply" is a state subject under the Seventh Schedule of the Indian constitution and under the provisions of the Eleventh Schedule, the responsibility has been devolved to the local self-governance institutions. Thus, it is quintessential to collaborate and bring in the Panchayati Raj Institutions to achieve the promise of providing access to safe drinking water in the rural areas. The said scheme (NRDWP) is a centrally sponsored scheme aimed at achieving this objective. It is a subcomponent of the Nirmal Bharat that focuses on the creation of rural infrastructure. This has led to an increased outlay and thus greater availability of resources for creating an ecosystem of sustainable infrastructure that can manage the task of ensuring safe access to drinking water in the rural areas.

3.6.9.26 Lohiya Swachh Bihar Abhiyan (LSBA)

Through LSBA, rural Bihar will be made ODF by 2 October 2019. An estimated 8332 Gram Panchayat will together be made ODF in 39 months, with a total of 1.6 crore toilets to be constructed under this program. The target is to make 6 Gram Panchayats per district ODF every month (on average). One Gram Panchayat will attain ODF in 90 days SBM (Gramin) and Lohiya Sachha Abhiyan will be implemented under LSBA in the state. It comprises construction of Individual Household Latrine (IHHL), Community Sanitary Complex, Solid Liquid Waste Management, and SBM (Gramin) is a centrally sponsored scheme with a 60:40 ratio (Center: State).

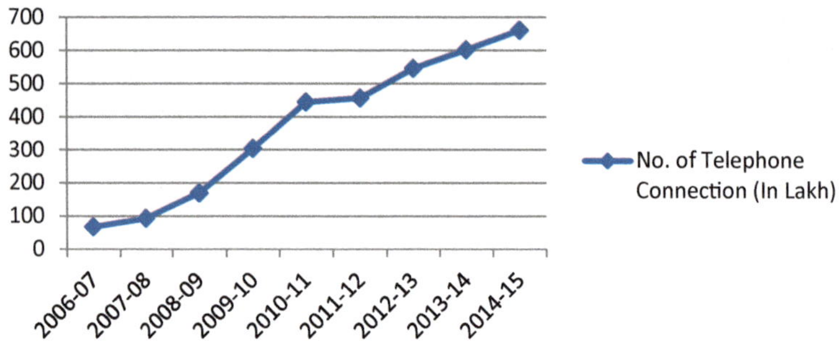

Fig. 3.17 Number of telephone connections, Bihar. *Source* TRAI (2015)

It covers toilet construction for BPL and identified APL. The rest of APL shall be covered through the LSY program of the Bihar government (Rural Development Department, Government of Bihar 2015).

3.6.9.27 MDG8

Goal 8 lays emphasis on 'developing a global partnership for development as it seeks to make available the benefits of new-age technologies, especially in the ICT domain, to all. India's growth in this field has been phenomenal as is evident through the rising profile of Indian telecom sector companies. Only China boasts of a better teledensity than India. With regard to Bihar, growth is mainly spearheaded by the private players who occupy 96% of the telephone connections in the state. The private connections in the state have risen 44 times from 14.18 lakhs to 632.56 lakhs between 2005 and 2014 (see Fig. 3.17).

3.7 Discussion

MDG1 deals with the eradication of extreme poverty in Bihar and it has done well in bringing down the poverty headcount ratio from 33.74% in the year 2011–12 to 49.4% in 2004–05, but still, 35.8 million people are living below the poverty line in Bihar. The major achievement had been seen in Goal-2, where the literacy rate rose from 47 in 2001 to 61.8% in 2011. Goal-3 is about meeting gender parity. Bihar is second in gender disparity in youth literacy rate. The ratio of female literacy rate to male literacy rate (15–24 years) is only 0.8 till 2011. In the last assembly election, 28 female members were elected to the legislative assembly of 243 members which is a mere 11.53%, and 4/75 women members are present in the legislative council of Bihar (Government of Bihar 2017). The under-five mortality rate (MDG4), which in 1992–93 was 128 per 1,000 live births, declined to 77 by 2010–11. Over the same period, the maternal mortality rate (MDG5) dropped from 531to 305 per 100,000 live births (SRS 2013). There has been some success in combating tuberculosis (TB) and malaria (MDG6) the annual new smear-positive case notification rate for TB increased from 20 per 100,000 populations in 2005 to 32 in 2011, while the number of reported malaria cases declined steeply from 2744 in 2006 to 496 in 2008. Cases reported for visceral leishmaniasis (kala-azar), a parasitic disease, dropped by 40% in the interval of 2007 to 2010 (Kumar et al. 2020). Table 3.14 present the district-wise trends in Sex Ratio, Child Sex Ratio, population Density, and Urbanization.

Table 3.14 Demographic trends of Bihar

District	Population (Lakh)		Sex ratio (Overall)		Sex ratio (Child)		Density		Urbanization	
	2001	2011	2001	2011	2001	2011	2001	2011	2001	2011
Patna Division										
Patna	47.2	58.4	873	897	923	909	1471	1803	41.6	43.1
Nalanda	23.7	28.8	914	922	941	931	1006	1220	14.9	15.9
Bhojpur	22.4	27.3	901	907	940	918	903	1136	13.9	14.3
Buxar	14.0	17.1	900	922	929	934	864	1003	9.2	9.6
Rohtas	24.5	29.6	910	918	952	931	636	763	13.3	14.5
Kaimur	12.9	16.3	901	920	942	942	383	488	3.3	4.0
Magadh Division										
Gaya	34.7	43.9	938	937	968	960	699	880	13.7	13.2
Jahanabad	9.2	11.3	927	922	915	922	963	1206	12.1	12.0
Arwal	5.9	7.0	931	928	917	940	–	1099	0.0	7.4
Nawada	18.1	22.2	946	939	978	945	726	880	7.7	9.7
Aurangabad	20.1	25.4	934	926	941	944	607	760	8.4	9.3
Saran Division										
Saran	966	954	949	926	1231	1493	1231	1493	9.2	8.9
Siwan	1031	988	933	940	1221	1495	1221	1495	5.5	5.5
Gopalganj	1001	1021	964	954	1057	1258	1057	1258	6.1	6.4
Tirhut Division										
W. Champaran	30.4	39.4	901	909	952	953	582	750	102	10.0
E. Champaran	39.4	51.0	896	902	935	933	991	1281	6.4	7.9
Muzaffarpur	37.5	48.0	921	921	927	915	1180	1506	9.3	9.9
Sitamarhi	26.8	34.2	892	892	924	930	1214	1491	5.7	5.6
Seohar	5.2	6.6	883	893	911	929	1161	1882	4.1	4.3
Vaishali	27.2	35.0	919	895	939	904	1332	1717	6.8	6.7

(continued)

Table 3.14 (continued)

District	Population (Lakh)		Sex ratio (Overall)		Sex ratio (Child)		Density		Urbanization	
	2001	2011	2001	2011	2001	2011	2001	2011	2001	2011
Darbhanga Division										
Darbhanga	33.0	39.4	914	911	913	931	14,442	1721	8.1	9.7
Madhubani	35.8	44.9	942	926	941	936	1020	1279	3.5	3.6
Samastipur	33.9	42.6	928	911	937	923	1175	1465	3.7	3.5
Munger Division										
Begusarai	23.5	29.7	911	895	947	919	1222	1540	4.6	19.2
Munger	11.4	13.7	872	876	916	922	800	958	27.9	27.8
Sheikhpura	5.3	6.4	920	930	964	940	762	902	15.6	17.1
Lakhisarai	8.0	10.0	919	902	954	920	652	815	14.7	14.3
Jamui	14.0	17.6	919	922	965	956	451	567	7.4	8.3
Khagaria	12.8	16.7	885	886	931	926	859	1115	5.9	5.2
Bhagalpur Division										
Bhagalpur	24.2	30.4	875	880	967	938	946	1180	18.7	19.8
Banka	16.1	20.3	909	907	964	943	533	672	3.5	3.5
Kosi Division										
Saharsa	15.1	19.0	911	906	910	933	885	1125	8.3	8.2
Supaul	17.3	22.3	921	929	927	944	724	919	5.1	4.7
Madhepura	15.3	20.0	916	911	930	930	853	1116	4.5	4.4
Purnia Division										
Purnia	25.4	32.6	916	921	968	954	787	1014	8.7	10.5
Kishanganj	13.0	16.9	934	950	946	971	687	898	10.0	9.5
Araria	21.6	28.1	914	921	963	957	751	992	6.2	6.0
Katihar	23.9	30.7	919	919	966	961	782	1004	9.2	9.9
Bihar	**830**		**919**	**918**	**942**	**935**	**880**	**1106**	**10.5**	**11.3**

Source Census (2011)

3.8 MDG's Challenges in Bihar

Flood and drought are the biggest challenges for Bihar that occur almost every year. In north Bihar, alone 76% of the population was hit by floods. In 2014, the state disaster management department declared 33 of the state's 38 districts drought-hit (Government of Bihar, World Bank and GFDRR 2008). Floods and drought can destroy governance and progress in Bihar. Infrastructure development is a big concern for the government. Bihar is in dire need of more infrastructures related to, health, road, and electricity. Unemployment is a major issue with widespread unemployment among youth. Earlier, migrants were mainly from poor households but today migration is among youngsters from all sections of society. Quality education is also a big challenge for good governance. The institutes are suffering from the scarcity of teachers, delays in academic sessions, and poor levels of research have become challenges.

3.9 Conclusion

Bihar is rising from the dust-biting state to the fastest growing economy in India. In the past two decades with its various policies in accordance with central programs and in line with MDGs, Bihar was able to reduce the proportion of people living below the poverty line. People are getting better health facilities and the state has achieved the target of universal enrollment ratio and gender enrolment ratio, maternal and infant mortality have reduced, and reservation in Panchayati Raj Institution has empowered the women further. More households have got connected through the clean water supply. Open defecation has been declining. Internet and mobile connectivity have improved. Connectivity has become cheaper and more prevalent due to the enhancement of Internet-based network.

Although the state expenditure in the social sector is increasing, Bihar stood at the bottom among other states of India. This resulted in minimal per capita development expenditure. The state needs to increase its investment in social infrastructure.

References

Adhikari SK, Bredenkamp C (2009) Moving towards an outcomes-oriented approach to nutrition program monitoring: the India ICDS Program. The World Bank, Washington, DC. Retrieved from: https://documents1.worldbank.org/curated/en/276711468267625639/pdf/494830WP0Box341enkampIndiaICDSFINAL.pdf

Adhikari A, Bhatia K (2010) Can we bank on banks? Econ Polit Wkly

Anand R, Tulin V, Kumar N (2014) India: defining and explaining inclusive growth and poverty reduction. IMF Working Paper

Annual Health Survey (2013) Annual Health Survey 2011–12. Fact Sheet Bihar. Registrar General & Census Commissioner, New Delhi, India. Retrieved From: http://statehealthsocietybihar.org/survey_reports/AHS-Bihar_Factsheet_2011-12.pdf

Benassy-Quere A, Bourguignon F, Dercon S, Estache A, Gunning JW, Kanbur R et al (2010) The millennium development goals: an assessment. In: Spence RK (ed), Equity and growth in a globalizing world, pp 17–40. The World Bank, Washinton DC. Available on: https://openknowledge.worldbank.org/bitstream/handle/10986/2458/548910PUB0EPI11C10Dislosed061312010.pdf?sequence=1&isAllowed=y

Bhalla SS (2013, July 24) The unimportance of NREGA. Indian Express. Available on: https://indianexpress.com/article/opinion/columns/the-unimportance-of-nrega/

Census (2011) Census of India 2011. Registrar General & Census Commissioner, India, New Delhi. Available on: https://censusindia.gov.in/2011Census/pes/Pesreport.pdf

Desai S, Vashishtha PS, Joshi OS (2015) Mahatma Gandhi National Rural Employment guarantee act: a catalyst for rural transformation. National Council of Applied Economic Research, New Delhi

Desjardins R, Schuller T, Feinstein L (2006) Understanding the social outcomes of learning. In Desjardins R, Schuller T, Feinstein L (eds) Measuring the effects of education on health and civic/social engagement, pp 11–15. OECD

Esteves T, Rao K, Sinha B, Roy SS, Rao B, Jha S et al (2013). Agricultural and livelihood vulnerability reduction through the MGNREGA. Econ Polit Wkly 94–103. Available on: http://www.jstor.org/stable/24477901

Gaur S, Rao NS (2020) Poverty measurment in India: a status update. Working Paper No. 1/2020. Ministry of Rural Development, Government of India, New Delhi. Retrieved From: https://rural.nic.in/sites/default/files/WorkingPaper_Poverty_DoRD_Sept_2020.pdf

Government of Bihar (2011) The Bihar state free and compulsory education of children rules, 2011. Human Resource Development Department, Government of Bihar, Patna. Available on: https://www.education.gov.in/sites/upload_files/mhrd/files/upload_document/bihar-rte-rules_english%20(1).pdf

Government of Bihar (2015) Report Card 2015. Bihar Vikas Mission, Governmnet of Bihar, Patna. Retrieved from: https://www.bvm.bihar.gov.in/reportcard

Government of Bihar (2016) Economic Survey 2015–16 of Government of Bihar. Ministry of Finance, Governmnet of Bihar, Patna. Available on: https://www.adriindia.org/centre/report_details/economic-survey-2015-16-of-government-of-bihar

Government of Bihar (2017) Economic Survey 2016-17. Ministry of Finance, Governmnet of Bihar, Patna. Retreive From: http://14.139.60.153/handle/123456789/12237

Bihar Kosi Flood (2008) Needs assessment report. World Bank & GFDRR, Government of Bihar, Patna

Government of India (2002) The Constitution (Eighty-sixth Amendment) Act, 2002. Government of India, New Delhi. Retrieved From: https://www.india.gov.in/my-government/constitution-india/amendments/constitution-india-eighty-sixth-amendment-act-2002

Government of India (2016) State-Wise percentage of population below poverty line by social groups, 2004-05. Department of Social Justice & Empowerment. Ministry of Social Justice & Empowerment. Government of India, New Delhi. Retreived From: https://socialjustice.gov.in/

Government of India (2022) Critical Evaluation of Mahatma Gandhi National Rural Employment Gurantee Act (MGNREGA). Tenth Report. Ministry of Rural Development. Government of India, New Delhi. Retrieved From: http://164.100.47.193/lsscommittee/Rural%20Development%20and%20Panchayati%20Raj/17_Rural_Development_and_Panchayati_Raj_20.pdf

Gragnolati M, Bredenkamp C, Gupta MD, Lee YK (2006) ICDS and Persistent Undernutrition: strategies to enhance the impact. Econ Polit Wkly, 1193–1201

IIPS (2021) National Family Health Survey (NFHS-4) 2015–16. India Fact Sheet. International Institute of Popuation Studies, Mumbai. Retrieved from: http://rchiips.org/nfhs/pdf/nfhs4/india.pdf

Khera R, Nayak N (2009) Women workers and perceptions of the National Rural Employment Guarantee Act in India. Gender pathway out of poverty: rural employment, pp 1–19. FAO-IFAD-ILO, Rome

Krishnan A (2018, November 11) Behind India's leap in ease of doing business. The Hindu. Available on: https://www.thehindu.com/business/Economy/behind-indias-leap-in-ease-of-doing-business/article25469900.ece

Kumar V, Mandal R, Das S, Kesari S, Dinesh DS, Pandey K, Das P et al. (2020) Kala-azar elimination in a highly-endemic district of Bihar, India: A success story. PLOS Negl Trop Dis, 1–27

Mattos FB, Dasgupta S (2017) MGNREGA, paid work and women's empowerment. International Labour Organization, Geneva. Available on: https://www.ilo.org/wcmsp5/groups/public/---ed_emp/documents/publication/wcms_613735.pdf

Mehrotra S, Sinha S (2017) Explaining falling female employment during a high growth period. Econ Polit Wkly 54–62

Ministry of Rural Development (2016) MGNREGA statistics. Ministry of Rural Development. Government of India, New Delhi

MOSPI (2017a) Millennium development goals—final country report. Central Statistics Office. Government of India, New Delhi. Available on: http://mospi.nic.in/sites/default/files/publication_reports/MDG_Final_Country_report_of_India_27nov17.pdf

MOSPI (2017b) Millennium development goals—final country report of India. Ministry of Statistics and Programme Implementation. Government of India, New Delhi

MoSPI (2017c) Millennium development goals India Country final report. Central Statistics Office, Ministry of Statistics and Programme Implementation, Government of India, New Delhi

MOSPI (2021) Sustainable development goals national indicators framework: progress report 2021. Ministry of Statistics and Programme Implementation. Government of India, New Delhi. Available on: http://mospi.nic.in/sites/default/files/publication_reports/SDG-NIF-Progress2021_March%2031.pdf

Mukul A (2015, September 23) Bihar school enrolment up, hits 94% at primary level. Times of India, pp. Retrieved From: https://timesofindia.indiatimes.com/home/education/news/bihar-school-enrolment-up-hits-94-at-primary-level/articleshow/49068112.cms

NACO (2015) India HIV Estimations 2015: Technical report. National AIDS Control Organisation & National Institute of Medical Statistics, ICMR. Ministry of Health & Family Welfare, Government of India, New Delhi

Naidu MV (2018) Address by Shri M. Venkaiah Naidu, honourable vice president of India at the international human rights conclave organized by the national human rights commission on the occasion of its Silver Jubilee Celebrations, in New Delhi on October 01, 2018. Government of India, New Delhi. Retrieved from: https://vicepresidentofindia.nic.in/speechesinterviews/address-shri-m-venkaiah-naidu-honourable-vice-president-india-international-human

Nelson RR (1956) A theory of the low-level equilibrium trap in underdeveloped economies. Am Econ Rev 894–908. Available on: http://www.jstor.org/stable/1811910

NSSO (2020) Annual report: periodic labor force survey (PLFS). Ministry of Statistics and Programme Implementation. Government of India, New Delhi

NSSO (2014) Participation of women in specified: activities along with domestic duties. NSS 68th Round. Ministry of Statistics and Programme Implementation. Government of India, New Delhi

NSSO (2015) NSS Report No. 566: status of education and vocational training in India. Ministry of Statistics & Programme Implementation, New Delhi. Available on: https://mospi.gov.in/documents/213904/301563//nss_report_no_566_21sep151602577933947.pdf/a5443bf8-4a37-ae08-f4f5-85f125bd1a2d

Panda B (2015) National rural employment guarantee scheme: development practice at the crossroads. Econ Polit Wkly, 50(23):126–131. http://www.jstor.org/stable/24482292

Pangariya A, Mukim M (2014) A comprehensive analysis of poverty in India. Asian Dev Rev 1–52. Available on: https://documents1.worldbank.org/curated/en/847861468257945118/pdf/946460v10JRN0B0O0900170adev0a000021.pdf

PIB (2015) India success story of a strong response to HIV/AIDS continues. Press Information Bureau. Government of India, New Delhi. Retrieved From: https://pib.gov.in/newsite/PrintRelease.aspx?relid=132173

Planning Commission (June 2014) Report of the expert group to review the methodology for measurement of poverty. Government of India, New Delhi. Retrieved From: https://niti.gov.in/planningcommission.gov.in/docs/reports/genrep/pov_rep0707.pdf

Poalnki P (2012, March 20) How UP and Bihar are skewing India's poverty estimates. First Post. Available on: https://www.firstpost.com/business/economy/how-up-and-bihar-are-skewing-indias-poverty-estimates-250226.html

Ravallion M, Datt G, Walle DV (1991) Quantifying absolute poverty in the developing world. Rev Income Wealth 345–361

Ravindranath NH, Murthy IK (2021) Mitigation co-benefits of carbon sequestration from MGNREGS in India. Plos One

RBI (2021) Table 10 : Per capita net state domestic product—state-wise (at constant prices). Reserve Bank of India, New Delhi. Available on: https://www.rbi.org.in/scripts/PublicationsView.aspx?id=20415
Registrar General of India (2011) Censul 2011. Registrar General of India. Government of India, New Delhi
Robert F, Kubiszewski I (2016) The UN sustainable development goals and the dynamics of well-being. Front Ecol Environ. https://doi.org/10.1002/fee.1231
Roberts K (1980) Price-independent welfare prescriptions. J Publ Econ 277–297. Available on: https://www.sciencedirect.com/science/article/abs/pii/0047272786900071
Ruskin J (1862) Unto this last: essays from the Cornhill Magazine 1860, reprinted as Unto This Last in 1862. Cornhill Magazine, London. Retrieved from: https://muff.uffs.net/skola/dejum/ruskin/texts/unto-this-last/unto_this_last.pdf
Samuelson PA, Swamy S (1974) Invariant economic index numbers and canonical duality: survey and synthesis. Am Econ Rev 566–593. Available on: http://www.jstor.org/stable/1813311
SECC (2011) SECC 2011 at a glance. Ministry of Rural Development, Government of India, New Delhi. Available on: https://secc.gov.in/getSeccDataSummaryNationalReport.htm
Sen A (1976) Poverty: an ordinal approach ro measurement. Econometrica 219–231
Sen A (1999a) Development as freedom. Knopf, New York
Sen A (1999b) Development as freedom: an India perspective. Indian J Indus Relat 157–169
Sen A (2001) Many faces of gender inequality. Frontline. Available on: https://www.sas.upenn.edu/~dludden/MANY%20FACES%20OF%20GENDER%20INEQUALITY.htm
Sen A (2003) Address at commission on population and development thirty-sixth session. United Nations, New York. Retrieved From: https://press.un.org/en/2003/pop859.doc.htm
Sen A (2005) The three R's of reform. Econ Polit Wkly 1971–1974. Available on: http://www.jstor.org/stable/4416610
Shah M (2007) Employment guarantee, civil society and Indian democracy. Econ Polit Wkly 43–51. Available on: http://www.jstor.org/stable/40276832
SRS (2013) Sample Registration System (SRS)-Bulletin 2012volume 48-II. Registrar General, New Delhi, India. Retrieved from: https://censusindia.gov.in/nada/index.php/catalog/42689
WHO (2015) Global Tuberculosis Report 2015. World Health Organization, Geneva. Retrieved from: https://apps.who.int/iris/bitstream/handle/10665/191102/9789241565059_eng.pdf?sequence=1
UNDP (2015) Building the Post-2015 development agenda: 2014 annual report. United Nation Development Programme, New York. Available on: https://www.undp.org/sites/g/files/zskgke326/files/publications/UNDP_Building%20the%20Post-2015%20Development%20Agenda_Final_2015.pdf
UNDP (2018) Human development report 2018. United Nation Development Programme and Oxford University Press, New York
UNICEF (2016) A snapshot of Bihar elementary education: district-wise statistical report. UNICEF BIHAR, Patna. Available on: https://www.bepcssa.in/en/A%20Snapshot%20of%20Bihar%20Elementary%20Education1.pdf
UNICEF (2017) A world free from child poverty: a guide to the tasks to achieve the vision. United Nation Children's Fund, New York
UNICEF (2020a) Maternal mortality in 2000: estimate developed by UNICEF and UNFPA. United Nations Children's Fund, New York
UNICEF (2020b) Universal child benefits: policy issues and options. UNICEF, New York. Available on : https://www.unicef.org/media/72916/file/UCB-ODI-UNICEF-Report-2020b.pdf
UNICEF India (2016) Children in India: India's gowth over the last two decades has contributed phenomenally to Global Human Development. Unitred Nation Children Fund, New Delhi
World Bank (2009) Doing Business in India 2009. The World Bank, Washington DC. Available on: https://openknowledge.worldbank.org/handle/10986/28088#:~:text=Doing%20business%20in%20India%202009%20is%20the%20first%20country%20specific,cities%2C%20and%20documents%20their%20progress

UNDP (1997) Human Development Report 1997. United Nation Devlopment Programme, New York
UNFPA (2016) Annual Report 2016. Millions of lives transformed. UNFPA. United Nations, New York. Retrieved from: https://www.unfpa.org/annual-report-2016

Chapter 4
Quantitative Analysis of Human Development in India

Abstract The eight Millennium Development Goals (MDGs) implementation in India is a case study from the developing world explaining the strength and weaknesses of this program. This chapter analyzes the impact of economic expansion on human development by examining the impact Gross Domestic Product (GDP) growth rate on the Human Development Index (HDI). It further focuses attention on the declining poverty but increasing inequality debate to assess the real achievement.

Keywords India · HDI · GDP · GNP · Poverty · Inequality

4.1 Introduction

The quantitative measurements of Millennium Development Goals (MDGs) are bound to be imperfect since these developments are multidimensional and qualitative. In particular, the measurement of socioeconomic development seems more controversial because countries differ considerably in both their institutional and social structures. In this study, Human Development Index (HDI) is taken as the economic parameter of MDG to determine development measures. The intention behind taking HDI as a parameter to determine the conditionalities is that HDI is an all-inclusive approach. In HDI, one finds all substantial requirements for determining an index of development. The determinants (i.e., life expectancy, education, and per capita income) cater to the MDG, SDG, and any development goals. The Impact of Gross Domestic Product (GDP) on HDI is designed to show the real development of the growing economy of a country. The impact of GDP on HDI is, therefore, expected to increase over time.

4.2 Human Development Index (HDI)

Human capital is considered the real wealth of the nation. The elementary objective of progress is to produce an empowering atmosphere for individuals to relish long, vigorous, and inspired lives. Technical deliberations on the means to attain human

U. K. Singh, *Human Development in Bihar, India*, Advances in Geographical and Environmental Sciences, https://doi.org/10.1007/978-981-19-3624-1_4

growth—and the usage of arithmetical aggregates to quantify national income and its growth—have at periods hidden the point that the main objective of growth is to help people. Following are the reasons for this:

- National income does not tell the real beneficiaries and the composition of income.
- There are certain achievements that do not get reflected in higher national income, like better health services, better access to knowledge, healthier working situations, refuge against lawbreaking and bodily viciousness, sufficient relaxation hours, etc.

"Current growth understanding has underscored the need for giving close consideration to the connection between economic development and human development for a variety of motives" (HDR 1990). They are:

- Many rapidly growing developing nations have realized that their tall GNP growth rates have been unsuccessful to decrease the socioeconomic deficiency of considerable sectors of their populace.
- Industrial countries are comprehending that great income is no defense against the fast spread of problems like drugs, intoxication, AIDS, penury, viciousness, and the failure of family relationships.
- Some little-income nations have established that it is possible to attain great levels of human growth if they competently use the obtainable means to enlarge rudimentary human competencies.
- Human growth exertions in many emerging nations have been sternly pressed by the economic catastrophe of the 1980s and the resultant modification programs.

Current growth knowledge is thus an influential cue that the enlargement of productivity and prosperity is only a means. The end of growth must be social happiness. So, the relation of the means to the final goal ought once to become the essential emphasis of growth investigation and planning.

Human growth is a procedure of increasing people's choices. The most serious is a lengthy and healthy life, enjoying a good standard of living and being educated. Supplementary choices include political autonomy, definite human privileges, and self-esteem. Income is a decent substitution for all other human selections since access to income allows the implementation of all other options. It is only partly true due to the following reasons (HDR 1990):

- Income is an end, not a means. It could be used for vital drugs or tranquilizer medications. The happiness of a culture is influenced by the uses of income not by the level of income itself.
- Nation experience establishes several cases of high levels of human growth at lesser income levels and meager levels of human growth at objectively high-income levels.
- Current income of a nation offers little supervision to its upcoming development scenarios. It has previously invested in its human capital, its possible revenue may be much more than what its present income level shows, and vice versa.

- Growing human problems in many rich countries demonstrate that high-income levels are no assurance of human progress.

The humble fact is that there is no reflex link between income growth and human growth. The main obsession of development analysis should be how such a connection can be shaped. Human growth denotes the procedure of spreading people's selections and the level of their attained comfort. It also supports differentiating clearly between two facets of human growth. One aspect is the formation of human competencies, such as improved health or knowledge. Another aspect is the use that people make of their learned competencies, for work or relaxation. This way of development differs from the conventional view in the following way:

- GNP growth is essential but not adequate for human growth. Human development may be deficient in some cultures despite rapid GNP development or high per capita income levels.
- Humans are more than capital goods for product assembly. They are the final ends and recipients of this procedure. Thus, the notion of human resource development does not capture the whole side of human development.
- Human well-being methods look at human beings as the recipients of the growth process. They highlight distributive strategies rather than manufacturing structures.
- The basic needs method focuses on the package of goods and services that disadvantaged people need like food, clothing, shelter, health care, water, etc. It emphasizes the delivery of these goods and services rather than human choices.

Human development carries together with the production and dissemination of merchandise and the growth and deployment of human competencies. It emphasizes choices. It is concerned not only with elementary needs gratification but also with human growth as an involved and active procedure. It applies to less advanced and highly advanced nations. This makes it conceivable to understand the complications of human life. Also, it is possible to appreciate many cultural, economic, social, and political alterations in people's lives.

According to Human Development Report (1990), the measurement of human development has 3 essential elements:

- Longevity—Life expectancy at birth is the indicator of this. Adequate nutrition and good health are closely related to life expectancy.
- Knowledge—Literacy is essential for learning and knowledge building. So, literacy is an indicator of knowledge.
- Decent living standards—Command over resources like land, credit, income, etc. can indicate decent living standards. However, due to a lack of data on some of these indicators, income and specifically per capita income could be taken as an indicator. However, the incidence of non-tradable goods and services and the alterations from exchange rate variances, tariffs, and taxes make per capita income data in nominal prices less valuable for worldwide comparisons. Such data can be enriched by using purchasing power-adjusted real GDP per capita figures. Further, the income indicator should reflect the shrinking returns to converting income into

human competencies. This aspect could be taken into consideration by using the logarithm of real GDP per capita for the income indicator.

The disparities in the overall population can be addressed by making distributional corrections. Such alterations are significant for income, which can rise to massive heights. The disparity possible in the case of life expectancy and literacy is much more restricted. An individual can be well-educated only once, and human life is limited.

Human Development Index (HDI) captures the overall sense of well-being of the people. HDI captures the three vital constituents of human life—longevity, knowledge, and basic income for a good living standard. Longevity and knowledge refer to the creation of human competencies, and income is a substitute measure for the choices people have in putting their abilities to use. The creation of HDI starts with a deprivation measure. For life expectancy, the target is 78 years, which is the maximum average life expectancy attained by any country. The literacy target is 100%. The income target is the logarithm of the average poverty line income of the better-off nations. It is stated in purchasing-power-adjusted international dollars.

4.3 HDI in India

India ranks 131 out of 188 countries with an HDI value of 0.624. It lies in the list of countries with medium human development. It falls along with countries such as Namibia (rank 126), Congo (rank 135), and Pakistan (rank 147). It is ranked third among the South Asian Association for Regional Cooperation (SAARC) countries. It ranks behind Sri Lanka (rank 73) and Maldives (rank 105). Both these SAARC countries fall in the high human development category. The top 3 countries in the world for human development are Norway (rank 1), Australia (rank 2), and Switzerland at rank 3 (HDR 2016).

In the 1990s, India was listed in the category of low human development. So, it is progress compared to the 1990s. Such progress is due to an increase in life expectancy and mean years of schooling in the past 25 years. Between the years 1990 and 2015, life expectancy at birth in India augmented by 10.4 years, expected years of schooling improved by 4.1 years, and mean years of schooling increased by 3.3 years. Gross National Income (GNI) per capita in India increased by about 223.4% from 1990 to 2015. This was largely due to India implementing market reforms, enticing investment, and dedicating more funds to social development, especially in sectors of health and education. The participation of women in the labor force of India is 26.8% and the Indian parliament is 12.2%. Indian population near multidimensional poverty is 18.2% and in severe multidimensional poverty is 27.8%.

The inequalities in India are evident from disparities across states. It can be seen in Table 4.1.

As can be seen from Table 4.1, Kerala has the highest human development in India with an HDI value of 0.712 whereas Bihar has the lowest with an HDI value

Table 4.1 HDI ranking, India

1	Kerala	0.712
2	Himachal Pradesh	0.670
3	Punjab	0.661
4	Maharashtra	0.666
5	Tamil Nadu	0.666
6	Haryana	0.661
7	Jammu & Kashmir	0.649
8	Gujarat	0.616
9	Karnataka	0.681
10	West Bengal	0.604
11	Andhra Pradesh	0.616
12	Assam	0.556
13	Rajasthan	0.577
14	Uttar Pradesh	0.542
15	Madhya Pradesh	0.557
16	Bihar	0.536
17	Odisha	0.557

Source HDR (2016)

of 0.536. The inequalities in human development, particularly in education, health, and living standards, stave off 27% of India's HDI score. So, the HDI score of 0.624 for India falls to 0.454 due to the inequality of distribution of the HDI dimension indices (Roche 2017).

4.4 Methodology

We use a simple OLS regression to estimate the extent to which financial deepening as measured by our monetization variable, affects growth in real GDP. In order to do this, a bunch of standard econometric tools are used which will be explained below.

Dummy variables taking values 0 and 1 help in assessing the categorical effect of variables, which might shift over time. The variables are categorized into mutually exclusive categories, like smokers and non-smokers.

In the regression done here, the country dummies show the additional effect a country has used. Given that there are two countries in the dataset, it is easy to use this technique to capture if one country has been more successful relative to the other in implementing institutional reforms in a way that has enhanced the effect of financial deepening on real GDP growth rate.

4.4.1 Coefficient of Determination

The R^2 coefficient of determination for regression is a statistical measure that is descriptive of how well the regression line borders on the real data points. The closer the value of R^2 is to 1, it indicates the regression line perfectly fits the data. Thus, the R^2 is a statistic that acquaints us with the goodness of fit for the model.

4.4.2 Time Trend

A whole new body of econometrics has grown up in the past 15 years or so to bargain with the rigidities of statistical analysis due to the time pattern. It is an unchallenging approach that adds the time trend to an OLS regression model. The properties of the intercept can be simply adjusted for a trend. For example, there are good prior reasons to suppose that a function is shifting, over time, due to factors other than those we are able to capture in the model it has specified. A repeated movement of the intercept can be represented. Further, it is suggestive that we could include a dummy for various shifts of the intercept. However, on falling short of degrees of freedom and hence being unable to estimate the equation. A settlement to overcome this is by surmising that the shift is the same each year. It implies a limitation that each year dummy shall have the exact same coefficient. For a linear equation, this presupposes an addition of a new "artificial" variable (which we can call T) to the equation. The series for the T variable necessitates the property of increasing the same constant amount each period. This is therefore a linear trend.

This gives us the model:

$$\text{in\_hdi} = \alpha + \beta * \text{in\_gdp}$$

assuming a simple bivariate linear model with a classical disturbance term.

4.5 Results

The following table summarizes the main results obtained from the regression (the appendix contains the entire output generated by STATA). The year variable captures the linear time trend in the evolution of real GDP. As already explained in a previous section, although not significant, this helps to extract a baseline growth in GDP and has been employed as a repressor by many in the literature. The second is the HDI variable used. Overall, for the combined dataset, development does seem to have a significant effect on the GDP growth rate.

The use of interacted dummies in standard regression analysis has already been explained. The interpretation here is that financial deepening is significantly more effective in promoting real GDP growth in India. The fact that this coefficient is statistically significant at the 99% confidence level validates my claim that institutional development is crucial in helping a country harness the full potential of its growth rate to promote development.

Estimate std. error *t* value Pr(> |*t*|)				
(Intercept)	0.441660	0.034980	12.626	4.33e−12***
ig	0.011777	0.005048	2.333	0.0284*

Residual standard error: 0.05643 on 24 degrees of freedom

Multiple R-squared (R^2): 0.1849	Adjusted R-squared: 0.1509
F-statistic: 5.443 on 1 and 24 DF, p-value: 0.02835	

Since the GDP of India is significant at 5%, it can be concluded that the HDI of India is explained by the GDP of India.

Also, R^2 15.09% of the variability in the HDI of India is explained by India's GDP.

p-value of the overall fit < 0.05, hence significant.

The econometric model that has been used fits well with the data as the overall fit is good. F test is basically used to test the hypothesis of all the coefficients in a regression equation being "0", that is, all the values are insignificant against the hypothesis of at least one of the values being significant. The lower the F value the better is the model. The value of the F statistic is used that proves that the model is an overall good fit for the data.

References

Human Development Report (1990) United Nations Development Programme, New York. Retrieved from https://hdr.undp.org/content/humandevelopment-report-1990

HDR (2016) Human development for everyone. Human Development Report. United Nations Development Programme. Retrieved from http://hdr.undp.org/sites/default/files/2016_human_development_report.pdf

Roche E (2017, March 22) India ranks 131 on Human Development Index, Norway No.1. Mint, pp. https://www.livemint.com/Politics/NcyY1Zr768TEl02yaRSh4M/India-ranks-131-on-global-Human-Development-Index-Norway-No.html

Chapter 5
Sustaining Development Process: Set of New Targets

Abstract There is considerable interest in the assessment of the present goals and in consideration of the future of development goals in the set of Sustainable development Goals. As the two-thirds mark for the achievement of these goals in 2015, this chapter focuses on the acceleration of progress, and on whether these goals are the right model for human development after 2015. It further understands the future improvements in goal setting benefits and identifies many common issues. This chapter analyses the crosscutting challenges as the basis to identify a set of principles for future goal development, after 2015. It focuses deliberately on challenges with the experiences of the MDGs so as to inform future goal setting in India.

Keywords India · MDGs · SDGs · Livelihood · Skill education · Health

5.1 Introduction

The mobilization of world leaders behind MDGs on the UN platform provided the most successful antipoverty programme in known history. To reaffirm the commitment to free the marginalized and the vulnerable from the abject of hunger and extreme poverty; provide equal opportunity to women, and protect our environment in the tune of shared prosperity. Member countries approved the set of 17 goals in the framework of sustainable development goals (SDGs) with the pledge to achieve these goals by 2030 (UN 2015a, b, c).

At the Rio conference of 2012, Member States agreed to implement a process to develop a set of goals and targets, i.e., Sustainable Development Goals (SDGs) based on the success of Millennium Development Goals and converges with the agenda 2030, i.e., "*no one will leave behind*". The SDGs replaced the MDGs which in September 2000 rallied the world around a common 15-year agenda to tackle the indignity of poverty.

For the first time, world leaders were vociferous about environmental loss at Stockholm Conference (1972), where leaders agreed on an urgent global response to the deteriorating environment. Brundtland Conference (1987) called for a "global agenda for change" and conceptualized the sustainable development goals with the aims of safeguarding the future while meeting the needs of today's generation. After

U. K. Singh, *Human Development in Bihar, India*, Advances in Geographical and Environmental Sciences, https://doi.org/10.1007/978-981-19-3624-1_5

two decades after Stockholm Conference, Rio Declaration (1992) advocated the need for environmental protection as essential for social and economic development. The Rio Conference was a significant milestone in making way for global goals. Johannesburg Summit (2002) amalgamated all the previous discussions and put forward a common path for human development through the lens of sustainable development. Rio + 20 Conference (2012) had identified the need for sustainable development at every point by integrating economic, social, and environmental determinants and recognized their synergy, to attain sustainable development in all dimensions of human development. The Member States agreed to adopt a process to develop a set of goals and targets, i.e., Sustainable Development Goals (SDGs) based on the success of the Millennium Development Goals and converges with the agenda 2030, i.e., "*no one will leave behind*". The SDGs replaced the MDGs which in September 2000 rallied the world around a common 15-year agenda to tackle the indignity of poverty.

International economic environment cooperation is needed to fuel sustainable development, particularly in the area of finance, technology transfer, debt management, and vital support of developing countries in raising the collective voice on the international forum to maintain sustainable progress and accelerate growth (UN 2015a, b, c). On the other side, consensus to maintain peace, security, and enabling human rights and liberty along with the right to development, and respect for cultural diversity, are essential concerns to achieving sustainable development and ensuring that sustainable development benefits all (UN 2002).

Making a sequel is not an easy task. Experience of MDGs helped in framing the future goals. Riding on the experience of MDGs the leaders agreed to give a call, to change the world by 2030. In the target year of MDGs, the world leaders met at the sustainable development summit (2015). The SDGs signify a transition in the future of human development. For the first time in known history, 193 countries agreed on a set of goals and targets, containing the issues dedicated to an equitable and sustainable way of living. The 17 goals are as follows:

5.1.1 Sustainable Development Goals

1. No Poverty,
2. Zero Hunger,
3. Good Health and Well-Being,
4. Quality Education,
5. Gender Equality,
6. Clean Water and Sanitation,
7. Affordable and Clean Energy,
8. Decent Work and Economic Growth,
9. Industry Innovation and Infrastructure,
10. Reduce Inequality,
11. Sustainable Cities and Communities,

12. Responsible for Consumption and Production,
13. Climate Action,
14. Life Below Water,
15. Life on Land,
16. Peace Justice and Strong Institutions,
17. Partnership for Goals.

Unlike MDGs, SDGs are universal goals framed to cater to the needs of developed, developing, and least developed countries of the world. The dedicated goals are more aggressive with a more comprehensive scope than MDGs. Inter alia, SDGs are contained divergent policy goals in the three dimensions—social, economic, and environmental. It suggests that the success of one goal depends on the success of others. It shows that the goals are exemplifying a common universal agenda and vision for progress in sustainable ways by creating space for all living creatures on the planet (Osborn et al. 2015).

5.2 Transition of Goals (From MDGs to SDGs)

Ban kin Moon expressed that the world in which we are living is witnessing rising inequality, diminishing trust, and intolerance and there is a need to support a new set of "global goals that have the potential to translate promises inked on paper to action on the ground" to realize the true sense of human development (UN 2015a, b, c).

Through the MDGs, the United Nations collaborated with the member states to provide a framework for developing policies and oversee aid programs designed to end poverty and uplift the poor people, achieve universal education, promote gender equality, reduce child mortality, reverse the spread of communicable diseases, ensure environmental sustainability, and provide a platform for universal development.

SDGs focus on the root causes of poverty and gender inequality as well as the holistic nature of development. Unlike MDGs, SDGs are meant for all either develop or developing since all countries are facing the problem of unemployment, lack of quality education, global warming, women empowerment, and climate change. Its goals reflect the paradigm shift in development to meet people's rights and needs, and the planet on sustainable development. The transition from MDGs to SDGs is framed in such a way that each goal on MDGs is further categorized into set goals, aligned with the agenda of human development (see Table 5.1).

The SDGs follow MDGs and expanded the development priority, which is aimed at ending poverty, hunger, disease, gender inequality, and increasing access to water and sanitation. The SDGs agenda look beyond the MDGs framework, addressing the universal wants for development, necessary for all.

Table 5.1 Transition of goals (from MDGs to SDGs)

MDG 1 (extreme poverty and hunger)	SDG 1 (no poverty)
	SDG 2 (zero hunger)
	SDG 8 (decent economic growth)
	SDG 10 (promote equality)
MDG 2 (universal primary education)	SDG 4 (quality education)
MDG 3 (gender equality and women empowerment)	SDG 5 (gender equality)
MDG 4 (infant mortality)	SDG 3 (health and well-being)
	SDG 2 (related targets)
MDG 5 (maternal health)	SDG 3 (health and well-being)
MDG 6 (communicable diseases)	SDG 3 (health and well-being)
MDG 7 (environment conservation)	SDG 6 (clean water and sanitation)
	SDG 7 (sustainable and clean energy)
	SDG 9 (sustainable innovation for industries and infrastructure)
	SDG 11 (sustainable cities)
	SDG 12 (responsible production and consumption)
	SDG 13 (climate sustainability)
	SDG 14 (sustainable use of water resource)
	SDG 8 (sustainable life on land)
MDG 8 (partnership for goals)	SDG 16 (partnership for development)
New entry	Peace, justice and institutional delivery

5.3 SDGs Baseline

At the end of 2015, eradicating poverty was one of the biggest challenges for mankind. Still, today around 700 million people live in extreme poverty and struggle to fulfill the basic necessities of life; health, education, and access to water and sanitation to name some. The majority of people living on less than $1.90 a day live in Southern Asia and sub-Saharan Africa and they account for about 70% of the global total of extremely poor people. Lower middle-income countries, including China, India, Indonesia, South Africa, and Nigeria, are home to about half of the global poor. However, this issue also affects developed countries. Right now, there are 30 million children growing up poor in the world's richest countries.

The final MDG report (2017) revealed the success story of India. The country has registered improvement in halving the poverty rates from the 1990 levels, in order to meet the target, the PHCR level has to be 23.9% by 2015. In 2011–12, the PHCR at all India levels was 21.9% which shows that India has already achieved the target well ahead of time.

SDG 1 aims to eradicate poverty in all its manifestations by 2030. The target of SDG 1 is to eradicate extreme poverty, measured as living expenses less than $1.25/day, from everywhere to everyone; minimize the proportion of men, women, and children living with multidimensional poverty; ensure coverage of social safety security for poor and vulnerable population.

Despite the unprecedented success in poverty reduction, children are still exposed to poverty. In developing countries, children constitute a third of the population, half of this population live in poverty. Child poverty is not limited to developing and least developed countries only, it is a global issue. The developed countries are not untouched. UNICEF surveyed children across 35 industrial countries and their finding suggests that an estimated 30 million children in OECD countries are living in poverty. The exact estimate of children living with MDP is lower than children living with monetary poverty. A higher number of MDP is expected among the children since the number of adults with MDP is 1.6 billion (UNICEF 2017).

The growth story of India in the past 25 years is inspiring. The country has contributed tremendously to global human development. When it comes to children, the color of success starts fading. India is a unique country in the world where the death rate of girls is higher than boys (UNICEF India 2016).

Goal 2 calls to end hunger and all forms of malnutrition by 2030. The targets include ending hunger in any form from anywhere, reducing the cases of stunting among the children under the age of and addressing the dietary requirement of lactating mothers; doubling the farm productivity; sustainable farming; forging investment in the agriculture sector, and liberalizing the agriculture markets.

The target of zero hunger can be achieved by sustainable food production systems and resilient agricultural practices, equal access to land, technology, and markets, and international cooperation on investments in infrastructure and technology to boost agricultural productivity. Ensuring that all people have access to safe, nutritious, and sufficient food all-year round is inextricably linked to poverty eradication and, as such, addressing undernutrition is indivisible from addressing poverty.

As per estimation, the world's food requirement will be doubled in the next years. The natural way to sustain agriculture is degrading because of climate change (World Bank 2008). Agriculture is a prerequisite to human development since agriculture, industries, and services are indivisible. It has a multidimensional impact on humanity, it eradicates hunger, improves income, and accelerates economic growth (Singh 2021).

Although India has achieved self-sufficiency in food grain, one-fourth of the country's population is food prone. The country is struggling to ensure availability of basic food requirement, the targets of SDG 2 focus on food with nutritious value.

Ensuring healthy lives and promoting well-being for all at all ages is important to building prosperous societies. However, despite progress in improving people's health and well-being in MDGs period, all the citizens are not able to access modern health care. Still today under-five mortality rates are very high; more than six million children still die before their fifth birthday each year, and only half of all women in developing regions have access to the health care they need. Epidemics like HIV/AIDS thrive the fear and discrimination limit people's ability to receive the services they need to live healthy and productive lives (UN 2017).

India is home to the second-largest population in the world. The country is recognized as an emerging economy. Notwithstanding, the progress during the period of MDG (2000–2015), life expectancy in India bounced from 63 to 68 years; infant mortality cases fell from 66 to 38 per 100 live births; maternal mortality improved from 364 to 174 per 100,000 live births (World Bank 2021), the country is still struggling to improve the accessibility of health services to poor. The challenges of the high mortality rate of maternal and neonatal cases. To realize SDGs, there is an urgent need for policy intervention in the health sector.

Education has multifaceted impact on human development. Literacy brings empowerment as it helps in employability, gives a sense of legal rights, facilitates mobility from deprivation, and with all these an individual can raise his/her voice for political rights.

SDG 4 advocates the need of education to leverage the human development. The goal call for knowledge and skills required to facilitate sustainable development. The goal is key to realizing the target of other SDGs. Through quality education, one can attain the basic necessities and live a decent life; this is the only weapon in the hand of marginalized and vulnerable that takes them out of poverty and deprivation.

The target of SDG 4 is to provide equal opportunity of education to all and focused on maintaining the supply of trained and qualified teachers in developing and least develop countries.

SDG 4 is steps ahead of MDG 2 which was limited to universal enrolment in primary education. Despite the success of MDG 2, millions of children were deprived of education, some are not attending school, and most of them who do attend are not learning. More than half of the youth and children lack proficiency in reading and solving mathematics. The situation is worse in sub-Saharan Africa (SSA) and Central and Southern Asia. Till 2015, 202 million (88%) in SSA and 241 million (84%) in Central and South Asian school students are found to be not proficient in reading while the numbers of students with poor mathematical skills is 193 million and 228 million, respectively (UN 2020a, b, c).

For India, the proportion of students with proficiency in reading is 47% in class III which decreased to 39% with upward mobility of classes (up to VII). The same declining trend is observed in understanding mathematics; 53% of students in class III and 40% up to VII standard (NIPUN BHARAT 2021).

Data presents a sorry picture of school education in India and the World. To attain the targets of SDG 4, immediate policy intervention is needed to strengthen the education sector of India, not only for economic development but also for human development too. India targets SDG 4 to equip their school-goers with better education which helps them in opening future avenues created by globalization.

SDG 5 is solely dedicated to women and aims to achieve equality and empowerment. Women are the fulcrum of society. It is inconceivable to achieve any target set for human development without the full participation of half the population. Gender equality is important because it has the potential to change economic outcomes by enhancing productivity. The economic empowerment of women translates into poverty reduction as women invest their earnings in their children and communities.

The target of SDG 5 is to free each and every woman and girl from any kind of discrimination against them so that no woman and girl experiences any type of violence or becomes prey to traffickers. The target further aims to stop child and forced marriage along with genital mutilation, granting the right of sexual and reproductive health, equal rights in property inheritance, enhance the use of ICT, and secure gender parity in all platforms of decision making.

Gender parity requires urgent attention because globally, 736 million, compromising 1 in 3 women, experienced physical and/or sexual violence in their life span since the age of 15 years. Around 10 million girls are at risk of child marriage. In lawmaking, the global average of women in national parliament is 25.6%, 36.3% in local governance and 28.2 women are holding managerial posts (UN 2021).

In India, in higher education 74% of women are not enrolled; only 65.46% of women are literate against 82.14% of men. The child sex ratio stands at 900 for 1000 men. In decision-making bodies, women constitute only 14% in the national parliament and below 10% in state assemblies. Women participation in the labor force is continuously decreasing, currently at 27.4% (UN 2020a, b, c).

Clean water and sanitation were recognized as human rights by United Nations in 2010. The decision was a landmark in the realization of human development because water is essential for the survival of humans and other living beings.

SDG 6 is called to ensure sustainable and affordable drinking water for all. The target aims to improve the access to sanitation and hygiene, end the practice of open defecation with a special focus on the requirement of women and vulnerable cohorts; make water free from contamination; integrate cross-border mechanisms for water management; protect the water ecosystem; international collaboration for and use artificial intelligence in aquaculture; support and incorporate tradition water management system.

Water Sanitation and Hygiene (WASH) are vital for human health and well-being. Safe WASH translates into human development by securing livelihood, promoting school attendance, maintaining dignity, and creating a healthy and resilient atmosphere for communities to live. Contaminated water has a significant impact on health as it damages vital body organs. It causes diarrhea, the second largest killer of children under the age of five.

An estimated 2 billion (26%) people across the world live in water-stressed countries. Nearly, 3.6 billion (46%) people do not have access to sanitation facilities and almost 2.3 billion (29%) are deprived of basic hygiene. By the end of 2015, 35% of wetland shrank.

Water with chemical contamination is another threat to humanity. Chemicals pollute water either in a natural way such as arsenic and fluoride or anthropogenic like nitrate. India is the world's worst groundwater contaminated with arsenic and fluoride. Most of the groundwater diseases in India are related to fluoride; 20 out of 29 states are fluoride affected. It affects nearly 40% (411 million) population scattered in 201 fluoride endemic districts of India and an estimated 66 million people in India are suffering from fluorosis of which 6 million are under the age of 14 (Chakraborti et al. 2016).

Although India has achieved the target to free the country from open defecation in 2019, through the world's biggest programme of behavior change, the country is struggling to provide 100% accessibility to toilets.

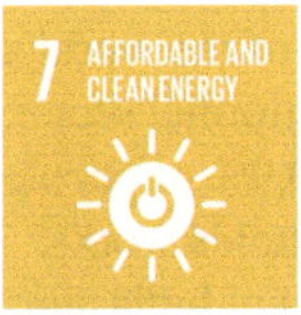

Fighting against climate change requires investment in clean and affordable energy. SDG 7 is committed to making sustainable clean energy affordable for everyone everywhere.

The target of SDG 7 is to expand the accessibility of modern energy services which are clean and affordable, add more weight to renewable sources in the world's energy mix, advancing energy efficiency by doubling the current rate, and international cooperation to enhance research in the field of sustainable and clean energy.

Energy along with water and land requires concerted effort for protection. Unfortunately, overexploitation destabilizes natural harmony and invites climate change. Optimum resource management is key to both, mitigation and adaptation resolve. A well-aligned policy approach is the need of time to deal with the shortcomings of energy, water, and land use.

Despite the improvement in access to electricity from 77.6% in 2000 to 85.3% in 2014, a population of 1.06 billion is still deprived of basic facilities of electricity. The urban–rural gap has narrowed with 96% of people in urban and 73% of people in rural areas had access to electricity. Clean fuel for combustion increased to 57 from 50% during the same period. The proportion of renewable energy in overall energy utilization increased from 17.5 to 18.3% between 2010 and 2014. Rapid industrialization demands a huge amount of energy. In India, almost 80% of energy consumption depends on 3 fossil fuels; coal 45.34%, crude oil 32.76%, and electricity 14.3%. The contribution of renewable sources (wind and solar) of energy in the final energy production of India increased from 6.7 GW in 2010 to 18.2 GW in 2019 (IEA 2021).

What is the meaning of decent work for most the people living across the world, International Labour Organization (ILO) emphasized decent work that is productive for men and women with conditions of equality, freedom, security, and human dignity.

SDG 8 is dedicated to decent work and economic growth. The targets are to maintain the annual 7% growth rate; promulgate high productivity by inducing technology; promote policies for decent jobs; accelerate economic growth without harming the environment; expand financial inclusion; galvanizing jobs for youth in collaboration with ILO.

In the emerging and developing economies of the world, a bigger proportion of people are engaged in the informal sector which is a threat to sustainable development. Informal sectors tend to remain isolated from the mainstream to protect themselves from the tax net translated into a lower tax base that hampers national economic growth. ILO (2018) estimation shows that 60% (2 billion) of the world's employed population (age above 15) is employed a least once in the informal sector. Further report revealed that, in India, the informal sector employed 80.9% of people while only 6.5% work in the formal sector and a mere 0.8% in households works.

With a focused program of employment generation, the Government of India aims to identify the mismatch in demand and supply of labor and trained them accordingly to the requirement of the market in accordance with the market.

SDG 9 is about sustainable future for everyone around the world. With the development going everywhere, urbanization is at a good pace, new cities are taking place. UNDP through SDG 9 aims to make sure that this happens in a sustainable way. Sustainable industries were supposed to produce clean sustainable products using renewable energy and also share innovative knowledge/technology equally with one another. Through the new innovation, anyone can sustain themselves in the competitive world with cutting-edge technology to sustain the humankind. To sustain the industries, we need sustainable infrastructure which provides a platform to build industries.

This target aims to develop and promote sustainable industrialization based on cutting-edge technology, new innovative ideas quality, reliable, sustainable, and resilient infrastructure, with a focus on affordable and equitable access for all. In the developing nations, SDG 9 targets small-scale industries and their integration into value chains and markets. By encouraging new innovative ideas and research to accelerate the economy vis-a-vis increase access to information technology and ensure the affordable access to the Internet in the least developed countries by 2020 (UNDP 2015). Organizing of unorganized sector through various programs in line with sustainable development helps India in achieving SDGs.

SDG 10 aims to raise the income of the people by 40% higher than the national average. It also targets to empower, promote, and provide equal opportunity to the social, economic, and political inclusion of all, irrespective of age, sex, disability, race, ethnicity, origin, religion, or economic or other statuses to reduce inequality. It also encourages ODA and FDI in accordance with their national plans and programs to strengthen the economy. The Government of India aims to reduce inequality across and within countries by 2030.

In the twenty-first century, urbanization is a prominent feature of developing nations. It is estimated that there may be a threefold increase in the total built-up area in developing countries between 2000 and 2030. Urbanization comes with economic growth and innovation throughout the world. On the other hand, urbanization has also contributed to environmental and socioeconomic challenges, including climate change, waste and pollution, congestion, and the rapid growth of slums.

SDG 11 promotes cities that are more inclusive, safe, resilient, and sustainable as they strive to achieve the Sustainable Development Goals. To ensure the availability of adequate, safe, and affordable housing for all, cities are built on the sustainable model with a sustainable transport system ensured for everyone notably special attention focused on women, children, persons with disabilities, and older persons. It aims to reduce the impact of natural disasters and substantially decrease the direct economic losses relative to GDP. SDG 11 focuses on the recommendation of the Sendai Framework for Disaster Risk Reduction 2015–2030, a holistic disaster risk management at all levels.

The Government of India is committed to better housing facilities for their citizens through measures like the "Smart City Mission" in India and "National Housing Policy: Supporting the People's Housing process", which are working to provide housing facilities to each people by 2030.

SDG 12's target is to implement the 10-year action plan on sustainable consumption and production and achieve sustainable management and efficient use of natural resources. Global waste is a major concern for the policy-makers, SDG 12 aims to reduce per capita global food waste by half and for better distribution system, strengthening the supply chain and management of post-harvest losses.

India, through programs and campaigns, promotes sustainable consumption and production by mainstreaming resource efficiency and sound management of waste. Governments can make significant achievements through awareness generation, education, and promotion of good habits and practices. As a signatory, India pledges to reduce the greenhouse gas (GHG) emissions intensity by 2030.

13 CLIMATE ACTION

For the developing world, climate change poses a serious threat to development and its widespread, unprecedented effects disproportionately burden the marginalized and vulnerable section of society. Goal 13 calls for urgent action not only to combat climate change and its impacts but also to build resilience in responding to climate-related hazards and natural disasters.

The Paris Agreement was a landmark deal, when 194 countries (including the EU and PRC), signed up to comprehensive pledges on the environment at a UN meeting in the French capital Paris in late 2015. The resolution aimed to constrain the global average temperature well below 2 °C above pre-industrial levels, the level beyond leads to global warming (UN 2015a, b, c). To limit the impact of climate change, the aim is to keep levels below 1.5 °C, above the pre-industrial level.

14 LIFE BELOW WATER

Natural waste management, with the breakdown and removal of waste and pollution, and their coastal ecosystems act as a shield to reduce damage from storms and other natural calamities. Its beaches become great places for tourism and recreation. Even more, it also helps in poverty reduction by increasing fish catches and income. As the small-scale fisheries engaged more and more women, it also imparts gender parity. The oceans are the home to dazzling creatures, ranging from single-celled organisms to the biggest animal ever to have lived on the Earth–the blue whale. They are also home to coral reefs, one of the most diverse ecosystems on the planet.

SDG 14 target is to prevent and significantly reduce marine pollution of all kinds by 2025 and to provide sustainable management to marine and coastal areas by 2020. It aims to limit the acidification of oceans and regulate overfishing and by 2020 prohibit certain forms of fishery subsidies. It also targets to enhance the conservation and sustainable use of oceans and their resources by implementing international laws in particular in its Articles 100–107 and 110, as in UNCLOS, which provides the legal structure to the conservation and sustainable use of oceans and their resources, as recalled in paragraph 158 of "The Future We Want".

In the course of their activities, individuals and commercial enterprises that use the ocean on a constant basis take decisions that affect the ocean. National governments and regional and global intergovernmental organizations all have their parts to play in regulating those activities.

SDG 15 ensures sustainable management of forests; wages a war against desertification; stops land degradation and biodiversity loss. Forests cover an estimated 31% of our earth's land area. It sustains the air we breathe, the water we drink, and the food we eat. Around 1.6 billion people depend on forests for their livelihood. Almost 75% of the world's poor are affected directly by land degradation. IPBES (2019) estimated forests accommodate more than 80% of terrestrial species of animals, plants, and insects. And of the 8,300 known species of animals, 8% are extinct and 22% are endangered.

SDG 15 target is to ensure the sustainable management of forests, wetlands, mountains, and drylands and to take immediate action to end poaching and trafficking of protected species of flora and fauna and address both demand and supply of illegal wildlife products by enhancing the global support for the same.

India has put in place an enabling legal mechanism under National Biodiversity Authority and a monitoring framework, National Biodiversity (NBT) Targets 28 with indicators for each of the targets. Synchronization of NBTs with SDG targets will contribute to the development of a monitoring framework post-2015 (Oommen and Laladhas 2016).

The United Nations Secretary-General António Guterres (2016) established a link between sustainable development and sustaining peace and told the General Assembly that the world needs a holistic framework to address the challenges ranging from sustainable development to human rights to maintain peace and manage prosperity.

SDG 16's target is to reduce all forms of violence, abuse, exploitation, trafficking, and all forms of violence, promote the rule of law at the national and international levels and ensure equal access to justice for all. It also aims to provide legal identity for all and enforce nondiscriminatory laws and policies for sustainable development.

SDG 16 speaks of peaceful and inclusive societies, access to justice for all, and effective and accountable institutions at all levels. Given India's initiatives in this regard, governments have been strengthening them while giving more attention to concerns specific to accountability in order to provide peace and justice and also ensure equal opportunity to participate in governance.

SDG 17 promotes partnership among the member countries to achieve SDGs. It facilitates the integration and provides an opportunity to learn from expertise of each other. It calls for partnership from each and every sector at each and every level.

The target of SDG 17 is to foster north–south along with south–south cooperation to take the world on the path of sustainable development. It facilitates international trade, promotes export from developing countries, and creates an environment for fair and equitable trade, benefiting all.

Through various multilateral and bilateral forum summits and many more, India is learning from the expertise of others in order to achieve the targets of SDGs.

5.3.1 Synergy Between SDGs

Many scholars examine the synergy and trade-off between the goals at length. The SDGs are interlinked, and the success of individual goals translates into the attainment of SDGs (Robert et al. 2016). Goals attainment relies on the maximization of synergy and minimization of the trade-off between the goals (Kroll et al. 2019). Smith and Olesen (2010) figured out the dual challenges of agriculture due to climate change, adaptation, and mitigation response. Climate change has the potential to halt the progress of poverty reduction and even reverse its success (Mathy and Blanchard 2016). Bue and Klasen (2013), argued that the success of one goal is likely to improve the other goals in frameworks of MDGs. Ibisch et al. (2016) findings suggest that technological advancement will degrade the natural environment. Nilsson et al. (2016) focus on local priorities, and coherence between the policies are key to the success of global goals. Pradhan et al. (2017) identified the synergy and trade-off between the goals by analyzing the SDG indicators for 227 countries, studies figured out the significant positive correlation between synergy and negative a trade-off. Engström et al. (2019) explored the impact of local policies on the global agenda and finds the intersection between the SDGs. Other studies established the synergy between SDG 7 (energy) and other SDGs (Nerini et al. 2017); Scherera et al. (2018) findings suggest that realization of SDG 1 (No poverty) translate into curbing inequality (SDG 10) likewise SDG 6 (Water and Sanitation), SDG 13 (climate), and SDG 15 (Life on Land). Allen et al. (2019) examine the 22 Arab countries and finds that prioritization of goals facilitates the SDGs. Neve and Sachs (2020) find a strong correlation between well-being and other SDGs, except SDG 12 (responsible consumption and production) and SDG 13 (climate action). There is a need to emphasize on localization of goals to build synergy and tradeoffs within the country's boundary and its impact across the world (Zhao et al. 2021).

5.4 Implementation of SDG in India

The success of the global agenda is based on the Sen (2005) 3R's, basic tools to initiate the development process in an economy; reach, range, and reason. The elaboration of 3R's is: (i) reach of goal to be achieved, (ii) means and methods used to realize the goal, and (iii) the justification for pursuit. SDGs are called to achieve each and every aspect of human development by 2030. It ensures peace, prosperity, and equality in a society within the ambit of sustainable development. The SDGs are not a legal agreement, but actually, an international obligation that depends on the moral vows and political determination aligned with the functioning of institutions. For the realization of SDGs depends on the localization of global goals, countries have to converge to their own policies for human development and develop a national framework to accomplish the goals. The SDGs implementation and attainment rely on countries' own schemes for sustainable development. Inter-Agency and Expert Groups (AIEG) on SDGs, identified and agreed on 230 indicators related to different goals, though the actual number of goals are 241 where 9 indicators reoccurred in 2 or 3 targets (UN 2016a, b). India as a signatory pledge to achieve the defined goals in the framework of SDGs. The country has implemented the SDGs in January 2016; national needs are prioritized in nationally defined indicators.

India has tasked the Ministry of Statistics and Programme Implantations (MOSPI) to develop a national framework based on local requirements. In 2018, after the first round, MOSPI had identified 306 national indicators in consultation with all stakeholders including Government Ministries and international agencies, in the span of 5 years NIF included 2 more indicators the bringing the total number to 308 (see Table 5.2).

Monitoring and evolution are key to the success of any goal. For the same purpose, the Government of India constituted a High-Level Steering Committee (HLSC), chaired by Chief Statistician and Secretary, MOSPI, the committee are represented by NITI Aayog, Ministry of Home Affair, Ministry of Health, Ministry of Environment, Forest and Climate Change, Ministry of Finance and MOSPI itself. To make the dynamics of the indicators, there is a technical advisory committee, constituted by the enactment of HLSC, empowered to add/delete indicators.

5.5 Convergence of SDGs with India's Priority

Based on the indicators given in NIF, India has characterized the national priority in social–economic, and environment which are profoundly indivisible dimensions

Table 5.2 National indicator framework (NIF)

Goals	Number of indicators		
	NIF 1.0	NIF 2.1	NIF 3.0
No poverty	19	21	22
Zero hunger	19	19	18
Good health and well-being	41	42	41
Quality education	20	19	20
Gender equality	29	29	29
Clean water and sanitation	19	16	16
Affordable and clean energy	5	5	5
Decent work and economic growth	40	32	33
Industry innovation and infrastructure	18	16	16
Reduce inequality	7	9	11
Sustainable cities and communities	16	15	15
Responsible consumption and production	17	15	15
Climate action	4	5	6
Life below water	13	11	11
Life on land	21	16	16
Peace justice and strong institutions	18	19	21
Partnership for goals	0	13	13
Total	306	302	308

Source MOSPI (2021)

prescribed in SDGs. The said dimensions are linked with the ongoing process. Growth and development are directly proportional, but more important is the redistribution of growth among the citizen to realize the development in the true sense, i.e., political, social, and cultural development. Traditional economists considered economic growth as a parameter for development (Meier 1976) but one cannot limit economic development with more and more productivity, it implies the transformation of the factor of production from labor intensive to automation.

5.6 Social Goals

The first five goals (SDG 1, SDG 2, SDG 3, SDG 4, and SDG 5), are recognized as social goals. An economy can achieve these goals by developing pro-poor policies such as expansion of the social safety net, improvement of the public healthcare system, improved investment in the education sector, and fixing the accessibility of institutions.

5.7 Economic Goals

Sustainable economic growth can be achieved by long-run policy framework, it's not an overnight work. It can be realized with political commitment and patience. Once an economy achieves the sustained high growth, it can channelize confidence, competitiveness, investment, job opportunity, resources, and help in expanding social security net and inclusiveness.

The goals related to economic aspects are SDG 8, SDG 9, SDG 10, SDG 12, and SDG 17. These goals can be achieved by committed leadership, effective policy measures, participation of stakeholders across the globe, local and international investment, innovation and entrepreneurship, world-class economic regulations for network industries and strategic waste management required for economic growth.

5.8 Environmental Goals

Environment is key to life support and major contributor in factor of production. People in emerging and least developing countries rely on natural resources to meet their basic necessities and livelihood. Concern for environment protection started galvanizing in the late twentieth century; a common consensus was developed to protect our natural environment against the negative impact of industrialization. It gives a clear message that benefits of high growth rate achieved on the cost of environment degradation are neither sustainable nor gives the sense of true development.

SDGs are interlinked. Purvis (2016) identified that the attainment of goals depends on strong synergy between all three aspects, social–economic–environmental. A subset of five goals SDG 6, SDG 7, SDG 11, SDG 13, and SDG 14 are committed to planet and prosperity, i.e., environment concern.

5.9 Peace Justice, Governance, and Partnership

SDG 16 and SDG 17 are focused on improving the accessibility to institutions and ensure the timely delivery. As common dictum about late delivery of institutions said "*justice delayed is justice denied*".

5.10 Discussion

Despite the robust economic growth of India in recent decades, social sectors are underdeveloped. Proper redistribution of economic growth improves living standards, consequently rise the demand in market and heralding the economic expansion (Sen and Dreze 2011). The argument is established in Chap. 4 that there is a strong correlation between the increase in GDP and HDI index. The indicators of the HDI are education, per capita income, and life expectancy. So, IT can be stated that increase in the GDP will lead to improvement of education, per capita income, and life expectancy. Once the poor are benefitted with high growth which is translated to rise in income, it helps in reducing poverty and improving other indicators of development such as health, education, water and sanitation, and environment too.

5.10.1 Social Safety Net

The developing concern of policy-makers is to develop such a programme that caters to the need of the poor population. Most of the countries have developed Social Safety Net (SSN) and Social Assistance (SA) programme as per their need. The programme aimed to take the people out of poverty. SDG 1 committed to end poverty in all its form by 2030, through SSN and SA basic necessities are ensured to the poorest of lower quantile and protect them from the negativity of externalities.

SSN and SA are instruments used by the government to enhance equity, promote resilience to lower the negative impact of economic and natural disaster, and help in building human capital through job creation. In the South Asia region, the share of public work is 25% of SSN. Within the region, Bangladesh and India are the countries with maximum investment in SSN. Public distribution programmes in India cost 1% of the country's GDP.

World Bank (2019) categorizes SSN policies into three ways:

1. Social Safety Net/Social Assistance (SSN/SA) Programmes are framed as non-contributory, to help poor people/households living in extreme poverty and struggling for the basic need of life. Tool of SSN/SA is cash transfer (conditional/unconditional), free meals in schools, educational fee waiver, and public work.

The government of India has enacted the Nation Food Security Act (NFSA) in 2013, the biggest ever food distribution programme in India. It covers almost 800 million people constituting 75% of all rural population and 50% of the urban population, through targeted Public Distribution Service (PDS), entitled to receive highly subsidized food grain for their survival. This programme cost more than one percent of India's GDP and occupied 70% of the total budget allocated for SSN.

NFSA has a multifaceted approach, one aspect is to make poor food secure and another is to feed nutrient food to pregnant women and lactating mothers under the umbrella cover of the Integrated Child Development System. Pregnant women are entitled to maternity benefit of more than or equal to Rs. 6000 as compensation for not working during pregnancy. Mid-day meal is there to serve to student up to the age of 14 years.

In this way, NFSA is pivotal in achieving the social goals of SDGs, it helps in achieving food security (SDG 1), potential to raise income (SDG 2), potential to reverse the MMR and IMR (SDG 3), help in promoting education through MMD (SDG 4), SDG 5 through ICDS network, by doing all these it reduces inequality (SDG 7).

Evidence suggests that NFSA is effective in poverty reduction in rural India. It is the main economic source for the poor in many states though it failed to improve the condition of the poor living in bigger state of India (Drèze et al. 2019).

Right to Education (RTE) is another SSN programme implemented in 2009 to provide free and compulsory education to all children in age group of 6–14 year. The financial burden of the programme is shared in concurrent way in between central government and state.

2. Social insurance programme (SIP) is a contributory step designed to help the person whose income is changed because of any reason (health, age, death of spouse, or natural calamity). Old age pension, maternity leave, survivor (widow) pension or health insurance are few examples.

Many developing countries like India, Bangladesh, Myanmar and Vietnam has implemented old-age pension scheme as measures of poverty alleviation.

Articles 41 and 42 of Indian Constitution ensure assistance to the individuals by the state in case of unemployment and securing humane condition such as maternity leave, respectively. On the same line, Government of India has launched National Social Assistance Programme (NSAP) in 1995 to provide the social assistance to the poor population in concurrent manner. Beneficiaries are identified on the basis of Below Poverty Line (BPL) census 2002. The programmes are

- National Maternity Benefit Scheme, implemented in 1999, benefits women above 19 years of age with Rs. 500/month during pregnancy, up to 2 live births, beneficiaries are poor households.
- National Old Age Pension Scheme (IGNOAPS), the scheme is meant for BPL individual, Rs. 200/month in the age group 60–79 and Rs. 500/month above 79 years.
- Indira Gandhi National Widow Pension Scheme, entitled Rs. 200/month to the widow of BPL community in the age group 40–60 years.
- Indira Gandhi National Disability Pension Scheme, disabled people in the age group 18–59 years, receive Rs. 200/month.
- National Family Benefits Programme, provides assistance of Rs. 10,000 on the demise of the primary bread earner of the family.
- Annapurna Programme, rolled in 2000, provides 10 kg of food grains to the individual above the age of 60, not covered in IGNOAPS.

3. Labor Market Programme (LMP) can be injected in both ways; contributory and non-contributory. Through dole it protects individuals who suffered income loss due to unemployment and imparts skill training to the person to meet the demand of the job market; both are parts of passive and active labor market policies, respectively.

Mahatma Gandhi National Rural Employment Guarantee (MGNREG) scheme covered 27% of the poorest quantile; the programme is observed as the largest job-providing scheme across the world. The scheme is designed to provide a minimum of 100 days of job to rural households. Programme was enacted in 2005, to raise the income of the rural population which helps in improving living standards.

MGNREGA works to uplift the downtrodden from the evil of poverty. Scholarships suggest that the said programme has a significant impact on poverty reduction in rural India. The programme performed better than any Union or State government employment scheme (Dreze et al. 2008), MGNREGA works resulted in bridging the rural–urban gap in social infrastructure and help in raising the farm productivity by strengthening irrigation infrastructure (Jacob 2008). Khera (2008) shows that MGNREGAscheme has improved the participation of farmers organization. Khera and Nayak (2009) carried a study on women involvement in MGNREGA through survey of Kaimur and Arraria (Bihar), Surguja (Chhattisgarh), Palamu (Jharkhand), Badwani and Sidhi (Madhya Pradesh), Dungarpur and Sirohi (Rajasthan) and Sitapur (Uttar Pradesh) they surveyed, figured that has potential to meet gender wage parity. Adhikari and Bhatia (2010) argued that the mechanism of direct cash transfer has reduced the problem of leakage and corruption at some extent. Women invest their earnings in to education of their children, which helps in building human capital (Afridi et al. 2016), overall rise in purchasing due to an increase in income expands the demand in the market (Sharma et al. 2016).

This passive labor market policy has the potential to achieve SDG 1, SDG 2, SDG 3, SDG 4, SDG 5, SDG 7, and have significant impact on other goals too.

The Government of India has implemented Skill India Programme as an active labor market policy to meet the mismatch in demand market and equip working-age population with skill education.

5.10.2 Skill Education

Another LMP is active labor policy, a contributory policy designed not for immediate gain but its contribution is observed in long run. It helps in building human capital and is an engine for sustainable growth.

Indian pattern of employment and productivity reveals important mismatches. Agriculture is the biggest job provider in India claiming 58% of India's workforce, the sector contributes 15% to the country's GDP. The agriculture sector suffers from subscale enterprises and overall low productivity. Baba Saheb Bhim Rao Ambedkar once said that the solution to capitalizing the surplus labor force in the agriculture sector is industrialization. Industry needs skilled labor, which is hard to find in India. It has become a big challenge ahead as it is estimated that only 4.69% of the total workforce in India has undergone formal skill training as compared to 68% in the UK, 75% in Germany, 52% in the USA, 80% in Japan, and 96% in South Korea (World Bank 2017).

The informal sector is a common feature of developing and emerging economies, where workers are employed without any protection of social security. The proportion of employment in the informal sector of India is almost 90%. Traditionally, the unemployment rate in India was never high or touched double-digit, but in 2016 it started to rise from 6%. ILO (2021) estimated that 30% (490 million) of Indian labor in India is under 30 years of age. The youth occupied the largest share of unemployment, nearly 83.3% of people in the age group of 15–29 is out of work. The country's youth population (age group 15–24 or 15–29) constitutes 37.5% of the country's population, burdened with 83.3% of unemployment. Lack of skill training is the main reason behind the unemployment in the youth of India.

Concern to convert the workforce into a skilled labor force, the country needs significant policy intervention. The Planning Commission report (1985) figured that through skill development country can sustain a GDP growth of 9% per annum, after that skill development programme became the apple of the eye of Indian policy-makers. India can reap the benefit of demographic dividend only by imparting quality education to the youth of the country, consequently attainment of decent jobs and better life and improvement in personal and professional life (Singh 2012).

Dedicated flagship programme to promote skill and training recognized the fact that a significant proportion of India's citizens are forced to take up unskilled work because they lack the quality education and skills training, a prerequisite to economically and professionally rewarding employment. For the first time in 2015, the Central Government of India has launched a Ministry to look after skill development programmes in India. The step is progressive in realizing the dream of work to every hand. The main concern being evolving pragmatic strategies to upgrade and enhance the relevant skill training and to improve access to skills for workers thus leveraging a cycle of higher productivity, higher employability, higher income levels, and faster and sustainable economic growth. Keeping its demand in view, on the occasion of the first-ever World Youth Skills Day on 15th July 2015, the Ministry of Skill Development and Entrepreneurship (MSDE) launched the Skill India Campaign. It aims to provide an umbrella cover to all skill activities being carried out within the country, to align them to common standards, and supply the trained workers as per the demand of the market.

The policy framework has been developed to accomplish the vision of Skill India by adhering to the objectives to attain sustainable development by empowering youth through the Skill India Mission. The framework outlines 11 major paradigms and enablers to achieve these objectives of skilling India are: (i) Institutional Training, (ii) Infrastructure, (iii) Convergence, (iv) Trainers, (v) Overseas Employment, (vi) Sustainable Livelihoods, (vii) Leveraging Public Infrastructure.

5.10.3 Health

Health is vital to economic development (Sen 1999a, b). Progress in income and education allow an individual/household to think about health and well-being. Good health and well-being ensure the active participation of an individual in the development process, within and beyond the health sector. A strong health system requires proper attention, investment, and people's participation. Evidence suggest that economic policies designed for sustainable growth are conducive to improving the health of citizens. World Bank report (1993) figured that during the 1980s, although CMR under 5 declined with the fall in income, the CMR in same age declined by double in the countries where income rose by 1%.

A person can experience good health and well-being on their own and its overall impact was felt on the economy's growth rate. Investment in health can be justified on economic grounds too. Health contributes to the economy in four ways: productivity improved with a smaller number of sick workers, improve the accessibility of workers to untapped natural resources, improve school attendance and ability to learn, and increased saving by halting medical expenses. On the other side, income's effect can be vital in improving accessibility to food, housing, and energy, and helps in population control by limiting family size.

Literature supports the linkage of health and economic growth. Bloom et al. (2004) findings suggest that the per capita GDP of a country can improve by nearly 4% with an added year in the life span of an individual due to better health facilities. Better income support improves accessibility to healthcare services, better dietary intake, less prone to disease, and higher accumulation of wealth (Thomas and Frankenberg 2002). Better life expectancy and higher remuneration for healthy workers can add good incentives to work; even the healthy workers take less time than unhealthy workers to complete the same assignment. The higher lifetime earnings and lower medical costs of health workers may lead to higher savings; such savings could be a conduit to progress and improving the living standard. Citizen's health has a positive and significant impact on real per capita income and overall economic growth (Sharma 2018). Deaton (2003) finds the correlation between health and inequality, he compares the rich and poor countries; rich and poor within the country, in rich countries people are healthier than people live in poor countries, people in rich countries have a healthier life than the poor country, at any moment of time, a wealthy person lives healthy life that the poor in the same country.

5.10.4 Challenges

Despite the reduction in poverty level in India, the proportion of poor expanded in rural areas due to the high growth rate of the population. The poor used to live in joint large families, relying on a narrow source of livelihood, which is the major constraint in breaking the poverty shield. Like other developing countries, nearly

70% of Indians live in rural area, in Bihar, almost 90% depend on agriculture for their livelihood, and most of them are marginal farmers and agricultural laborers. The small size of land holding lower productivity and wage, consequently restricting the mobility from poor to prosperous. Notwithstanding the decrease in the evidence of IMR, CBR, and CDR across the country, the declining rate was slower in the State of Bihar and Uttar Pradesh because of the poor health infrastructure.

In India, at the national level and in Bihar at the state level, the bigger challenge for the policy-making is to realize the benefits of population dividends. At present, India has 31% of its population under the age of 14, the corresponding figure is 40.1% for Bihar; 4 out of every 10 people in Bihar is under the age of 14 and that is remarkably a young state in India. This is the highest of such proportion for any state; it is possibly even the world's highest. So, while the rest of the country has lowered the proportion of its population that is under 14, for Bihar, this proportion has increased by over 10% since 2001. This makes it the only state that has added more persons under the age of 14 than over the age of 60. The working-age group population is sizeable and needs to get involved in the state's workforce as employment is the main source of livelihood for most of the people State government has to share the responsibility of imparting skill development programmes to its citizen so that they taste the fruit of global economy.

Well-skilled labor is productive and hence contributes toward robust economic growth. Lack of access to employment lowers self-esteem and leads to denial of basic needs of the individual and the family, and can lead to social instability. Bihar is suffering from a twin whammy of unemployment and poverty. Close to half of Bihar's young people depend on farming for a living. In comparison to other states of India, educated and qualified youngsters are more unemployed in Bihar than youth who are illiterate or with low education.

Bihar envisages venturing on a voyage of accomplishing a utopian state in the skill development space. State government's work on call of the hour is to create a holistic atmosphere for skill development, an umbrella cover for the various stakeholders, in order to sustain the economy and benefit the society at large.

With the implementation of SDGs in January 2016, a great deal of energy is being spent deliberating on what the new goals should be, which are the set of 17 goals and 169 targets. The SDGs are more ambitious than the MDGs, covering a broad range of interlinked issues, from economic growth to social issues to climate change. The implementation of SDGs requires every country to judiciously prioritize, and adapt the goals and targets in accordance with local challenges, capacities, and resources available.

The first level of estimates indicates a financial shortfall of INR 533 lakh crores (USD 8.5 trillion) over the mandated 15 years for achieving SDGs (TARA 2015). Per year, on average, this works out to INR 36 lakh crores or USD 565 billion. (Note that this is only the gap in finance to achieve the SDGs, not the overall financial requirement.) This is a minimalist estimate and the actual amounts are likely to be much higher (Technology and Action for Rural Advancement 2015). MDG Report 2014 showed that despite high economic growth, in 2010, one-third of the world's 1.2 billion extremely poor lived in India alone. Given these constraints, it is likely that domestic revenues aside, private finance could be a crucial source for financing the SDGs.

5.11 Conclusion

Sustainable Development Goals were born at the United Nations Conference on sustainable development in Brazil in 2012. These goals are based on the success of the Millennium Development Goals, to finish the unfinished targets. These goals are meant for all the countries, either higher income or lower income. The first five goals of SDGs (No Poverty, Zero Hunger, Good Health and Well-Being, Quality Education, and Gender Equality) are based on MDG 1 (Eradicate extreme poverty and hunger). SDG 8, SDG 9, and SDG 10 correspond to denote the living condition of the people by means of decent and new innovative ideas and sustainable economic growth. SDG 6, SDG 7 SDG 11, SDG 12, SDG 13, SDG 14, SDG 15 SDG 16, SDG 17 deal with climate change. These goals are indivisible and balance the economic, social, and environmental dimensions. These goals can be realized only by fostering gender equality, and not otherwise. Sector-specific assessment of progress toward MDGs identifies health problems, lack of quality education and income inequality are the major issues towards the achievement of MDGs. SDG funding, corruption, and the impact of climate change are the prominent challenges to attaining the post-2015 goals.

References

Abrahamsen R (2004) The power of partnerships in global governance. Third World Q 1453–1467

Adhikari A, Bhatia K (2010) Can we bank on banks? Econ Polit Wkly

Afridi F, Mukhopadhyay A, Sahoo S (2016) Female labour force participation and child education in India: the evidemce from national rural employment gurantee scheme. IZA J Lab Dev 1–27

Ahluwalia MS (2020) Economic performance of states in post-reforms period. Econ Polit Wkly 1637–1648

Alkire S, Roche JM (2013) How multidimensional poverty went down: dynamics and comparisons. University of Oxford, Oxford. Available on: https://www.ophi.org.uk/wp-content/uploads/How-Multidimensional-Poverty-went-down.pdf

Allen C, Metternicht G, Wiedmann T (2019) Prioritising SDG targets: assessing baselines, gaps and interlinkages. Sustain Sci 14:421–438. Available on: https://link.springer.com/article/10.1007%2Fs11625-018-0596-8

Anand R, Tulin V, Kumar N (2014) India: defining and explaining inclusive growth and poverty reduction. IMF working paper

Bank W (2019) The state of social safety nets: 2018. The World Bank Group, Washington, DC

Benassy-Quere A, Bourguignon F, Dercon S, Estache A, Gunning JW, Kanbur R et al (2010) The millennium development goals: an assessment. In: Spence RK (ed) Equity and growth in a globalizing world. The World Bank, Washington, DC, pp 17–40. Available on: https://openknowledge.worldbank.org/bitstream/handle/10986/2458/548910PUB0EPI11C10Dislosed061312010.pdf?sequence=1&isAllowed=y

Bhalla SS (2013) The unimportance of NREGA. Indian Express, July 24. Available on: https://indianexpress.com/article/opinion/columns/the-unimportance-of-nrega/
Bloom DE, Canning D, Jamison DT (2004) Helath, wealth and wellfare. Financ Dev 10–15
Bue MC, Klasen S (2013) Identifying synergies and complementarities between MDGs: results from cluster analysis. Soc Indic Res 647–670. Availble on: https://link.springer.com/article/10.1007%2Fs11205-013-0294-y
Census (2011) Census of India 2011. Registrar General & Census Commissioner, India, New Delhi. Available on: https://censusindia.gov.in/2011Census/pes/Pesreport.pdf
Chakraborti D, Rahman MM, Chatterjee A, Das D, Das B, Nayak B, Roy MM (2016) Fate of over 480 million inhabitants living in arsenic and fluoride endemic Indian districts: magnitude, health, socio-economic effects and mitigation approaches. J Trace Elem Med Biol Organ Soc Miner Trace Elem (GMS) 33–45. Available on: https://doi.org/10.1016/j.jtemb.2016.05.001
Deaton A (2003) Health, inequality, and economic development. J Econ Lit 113–158
Desai S, Vashishtha PS, Joshi OS (2015) Mahatma Gandhi national rural employment guarantee act: a catalyst for rural transformation. National Council of Applied Economic Research, New Delhi
Desjardins R, Schuller T, Feinstein L (2006) Understanding the social outcomes of learning. In: Desjardins R, Schuller T, Feinstein L (eds) Measuring the effects of education on health and civic/social engagement. OECD, pp 11–15
Dreze J (2021) Where is the strategy for dealing with learning loss during Covid? The Indian Express, 17 Sept. Available on: https://indianexpress.com/article/opinion/columns/where-is-the-strategy-for-dealing-with-learning-loss-during-covid-7511344/
Dreze J, Dev N, Khera R (2008) Corruption in NREGA: myth and reality. The Hindu, Jan 22
Drèze J, Gupta P, Khera R, Pimenta I (2019) India's public distribution system after the food security act. Econ Polit Wkly 36–47
Engström RE, Destouni G, Howells M, Ramaswamy V, Rogner H, Bazilian M (2019) Cross-scale water and land impacts of local climate and energy policy—a local Swedish analysis of selected SDG interactions. Sustainability 242–247. https://doi.org/10.3390/su11071847
Esteves T, Rao K, Sinha B, Roy SS, Rao B, Jha S et al (2013) Agricultural and livelihood vulnerability reduction through the MGNREGA. Econ Polit Wkly 94–103. Available on: http://www.jstor.org/stable/24477901
FAO (2020) The state of food security and nutrition in the world. Food and Agriculture Organization, Rome
Ferreira FH, Lugo MA (2013) Multidimensional poverty analysis: looking for a middle ground. World Bank Res Obs 220–235. Available on: https://doi.org/10.1093/wbro/lks013
Fukuda-Parr S (2004) Millennium development goals: why they matter. Glob Gov 10(4):395–402
Global Hunger Index Report 2019 (2019) Cocern Worldwide, Dublin
Government of Bihar (2011) The Bihar state free and compulsory education of children rules, 2011. Human Resource Development Department, Government of Bihar, Patna. Available on: https://www.education.gov.in/sites/upload_files/mhrd/files/upload_document/bihar-rte-rules_english%20(1).pdf
Government of Bihar (2016) Economic survey 2015–16 of Government of Bihar. Ministry of Finance, Governmnet of Bihar, Patna. Available on: https://www.adriindia.org/centre/report_details/economic-survey-2015-16-of-government-of-bihar
Haines A, Smith KR, Anderson D, Epstein PR, McMichael AJ, Roberts I et al (2007) Policies for accelerating access to clean energy, improving. The Lancet 370:1264–1281
Hazarika I (2012) India at the crossroads of millennium development goals 4 and 5. Asia Pac J Public Health 450–463
Hirschman AO (1958) Strategy of economic development. Yale University Press, Yale
Hulme D, Wilkinson R (2012) The millennium development goals and beyond: global development after 2015. Routledge

Ibisch PL, Hoffmann MT, Kreft S, Pe'er G, Kati V, Biber-Freudenberger L et al (2016) A global map of roadless areas and their conservation status. Science 354(6318):1423–1427. Available on: https://doi.org/10.1126/science.aaf7166

IEA (2021) India energy outlook 2021: world energy outlook special report. International Energy Agency, Paris. Available on: https://iea.blob.core.windows.net/assets/1de6d91e-e23f-4e02-b1fb-51fdd6283b22/India_Energy_Outlook_2021.pdf

ILO (2018) Women and men in informal economy: a statistical picture, 3rd edn. International Labour Organization, Geneva. Available on: https://www.ilo.org/wcmsp5/groups/public/---dgreports/---dcomm/documents/publication/wcms_626831.pdf

ILO (2021) Young persons not in employment and education (NEET) in India: 2000–2019. International Labour Organization, New Delhi. Available on: https://www.ilo.org/wcmsp5/groups/public/---asia/---ro-bangkok/---sro-new_delhi/documents/publication/wcms_789243.pdf

IPBES (2019) The global assessment report on Biodiversity and Ecosystem Services: summary for Policymakers. Bonn. IPBES secretariat. Retrieved From: https://ipbes.net/sites/default/files/inline/files/ipbes_global_assessment_report_summary_for_policymakers.pdf

ITU (2021) Individuals using internet (from any location by gender and urban/rural location %). International Telecommunication Union, Geneva. Available on: https://www.itu.int/en/ITU-D/Statistics/Pages/stat/default.aspx

Jacob N (2008) The impact of NREGA on rural-urban migration: field survey of Villupuram District, Tamil Nadu. CCS working paper no 202. Centre for Civil Society

Kadiyala S, Barnett T (2004) Econ Polit Wkly 39(19):1888–1892

Kapoor A, Sahoo RL (2004) Agricultutralists and the moneylender: a study of rural borrowing in Bihar and Orissa 1912–1936. In: Proceedings of the Indian history congress. Indian History Congress, pp 491–500. Available on: https://www.jstor.org/stable/44144764

Kaushal KK, Melkani VK (2005) India: achieving the millennium development goals through non-timber forest products. Int Forest Rev 128–134

Khera R (2008) Employment Guarantee Act. Econ Polit Wkly, August 30, 8–10

Khera R, Nayak N (2009) Women workers and perceptions of the national rural employment guarantee act in India. In: Gender pathway out of poverty: rural employment. FAO-IFAD-ILO, Rome, pp 1–19

Kline P, Moretti E (2013) People, places, and public policy: some simple welfare economics of local economic development programs. Working paper 19659. NBER. Available on: https://www.nber.org/system/files/working_papers/w19659/w19659.pdf

Krishnan A (2018) Behind India's leap in ease of doing business. The Hindu, 11 Nov. Available on: https://www.thehindu.com/business/Economy/behind-indias-leap-in-ease-of-doing-business/article25469900.ece

Kroll C, Warchold A, Pradhan P (2019) Sustainable development goals (SDGs): are we successful in turning trade-offs into synergies? Palgrave Commun 5. Available on: https://doi.org/10.1057/s41599-019-0335-5

Kumar A, Valecha N, Jain T, Dash AP (2007) Burden of malaria in India: retrospective and prospective view. Am J Trop Med Hygiene 69–78

Lahiry S (2014) India's millennium development goals. World Aff J Int Issues 18(3):140–149

Loiwal M (2020) India Today, 31 Mar

Manning R (2009) Using indicators to encourage development: lesson from millennium development goals. DIIS. Danish Institute for International Studies, Copenhagen

Mathy S, Blanchard O (2016) Proposal for a poverty-adaptation-mitigation window within the green climate fund. Clim Policy 752–767. Available on: https://doi.org/10.1080/14693062.2015.1050348

Mattos FB, Dasgupta S (2017) MGNREGA, paid work and women's empowerment. International Labour Organization, Geneva. Available on: https://www.ilo.org/wcmsp5/groups/public/---ed_emp/documents/publication/wcms_613735.pdf

McMichael AJ, Friel S, Nyong A, Corvalan C (2008) Global environmental change and health: impacts, inequalities, and the health sector. BMJ 191–194

Mehrotra S, Sinha S (2017) Explaining falling female employment during a high growth period. Econ Polit Wkly 54–62
Meier GM (1976) Leading issues in economic growth. Oxford University Press, New York
Ministry of Education (2021) Initiatives by the school education sector in 2020–21 for continuing teaching and learning. Minstry of Education Government of India, New Delhi. Available on: https://dsel.education.gov.in/sites/default/files/update/covid_initiatives.pdf
Ministry of Rural Development (2011) Ministry of Rural Development, Government of India, New Delhi
mint (2016) Bihar extends 35% reservation for women in all govt jobs. mint, Jan 2016. Available on: https://www.livemint.com/Politics/jNlMwoB5KuaHUZX767EbeI/Bihar-extends-35-reservation-for-women-in-all-govt-jobs.html
Mohanty SK (2012) Multiple deprivations and maternal care in India. Int Perspect Sex Reprod Health 38(1):6–14
MOSPI (2017a) Millennium development goals—final country report. Central Statistics Office, Government of India, New Delhi. Available on: http://mospi.nic.in/sites/default/files/publication_reports/MDG_Final_Country_report_of_India_27nov17.pdf
MOSPI (2017b) Millennium development goals—final country report of India. Ministry of Statistics and Programme Implementation, Government of India, New Delhi
MOSPI (2021) Sustainable development goals national indicators framework: progress report 2021. Ministry of Statistics and Programme Implementation, Government of India, New Delhi. Available on: http://mospi.nic.in/sites/default/files/publication_reports/SDG-NIF-Progress2021_March%2031.pdf
Moyo S, Wright JA, Ndamba J, Gundry S (2004) Realising the maximum health benefits from water quality improvements in the home: a case from Zaka district, Zimbabwe. Phys Chem Earth 1295–1299
Nelson RR (1956) A theory of the low-level equilibrium trap in underdeveloped economies. Am Econ Rev 894–908. Available on: http://www.jstor.org/stable/1811910
Nerini FF, Julia Tomei LS, Bisaga I, Parikh P, Black M, Borrion A et al (2017) Mapping synergies and trade-offs between energy and the sustainable development goals. Nat Energy 10–15. Available on: https://www.nature.com/articles/s41560-017-0036-5
Neve J-ED, Sachs JD (2020) The SDGs and human well-being: a global analysis of synergies, trade-offs, and regional differences. Sci Rep 1–12. Available on: https://www.nature.com/articles/s41598-020-71916-9
Nilsson M, Griggs D, Visbeck M (2016) Policy: map the interactions between sustainable development goals. Nature 320–322. Avaoilable on: https://www.nature.com/articles/s41599-019-0335-5#ref-CR15
NIPUN BHARAT (2021) National initiative for proficiency in reading with understanding and numeracy: guideline for implementation. Department of School Education & Literacy, Ministry of Education, Government of India, New Delhi. Available on: https://www.education.gov.in/sites/upload_files/mhrd/files/nipun_bharat_eng1.pdf
NSSO (2014) Participation of women in specified: activities along with domestic duties. NSS 68th round. Ministry of Statistics and Programme Implementation, Government of India, New Delhi
NSSO (2015) NSS report no. 566: status of education and vocational training in India. Ministry of Statistics & Programme Implementation, New Delhi. Available on: https://mospi.gov.in/documents/213904/301563//nss_report_no_566_21sep151602577933947.pdf/a5443bf8-4a37-ae08-f4f5-85f125bd1a2d
NSSO (2020) Annual report: periodic labor force survey (PLFS). Ministry of Statistics and Programme Implementation, Government of India, New Delhi
Oommen OV, Laladhas KP (2016) Sustainable management and use of terrestrial ecosystem. In: RIS, India and sustainable development goals: the way forward. Research and Information System for Developing Countries (RIS), New Delhi, pp 181–194

Osborn D, Cutter A, Ullah F (2015) Universal sustainable development goals: understanding the transformational challenges for developed countries. United Nations, New York. Available on: https://sustainabledevelopment.un.org/content/documents/1684SF_-_SDG_Universality_Report_-_May_2015.pdf

Pangariya A, Mukim M (2014) A comprehensive analysis of poverty in India. Asian Dev Rev 1–52. Available on: https://documents1.worldbank.org/curated/en/847861468257945118/pdf/946460v10JRN0B0O0900170adev0a000021.pdf

Piot P, Greener R, Russell S (2007) Squaring the circle: AIDS, poverty, and human development. PLoS Med 4(10):e314, 1571–1575

Planning Commission (1985) Seventh five year plan: 1985–1990. Planning Commission of India, New Delhi

Poalnki P (2012) How UP and Bihar are skewing India's poverty estimates. First Post, 20 Mar. Available on: https://www.firstpost.com/business/economy/how-up-and-bihar-are-skewing-indias-poverty-estimates-250226.html

Poku N, Whitman J (2011) The millennium development goals: challenges, prospects and opportunities. Third World Q 3–8

Pradhan P, Luís Costa DR, Kropp JP (2017) A systematic study of sustainable development goal (SDG) interactions. Earth's Future 1169–1179. Available on: https://doi.org/10.1002/2017EF000632

Purvis K (2016) 12 ways environment and development sectors can collaborate to meet the SDGs. The Guardian, 31 Oct

Radkar A, Parasuraman S (2007) Maternal deaths in India: an exploration. Econ Polit Wkly 3259–3263

Ravallion M, Chen S (2003) Measuring pro-poor growth. Econ Lett 93–99

Ravallion M, Datt G, Walle DV (1991) Quantifying absolute poverty in the developing world. Rev Income Wealth 345–361

Ravindranath NH, Murthy IK (2021) Mitigation co-benefits of carbon sequestration from MGNREGS in India. PLoS ONE

RBI (2021) Table 10: per capita net state domestic product—state-wise (at constant prices). Reserve Bank of India, New Delhi. Available on: https://www.rbi.org.in/scripts/PublicationsView.aspx?id=20415

Registrar General of India (2011) Censul 2011. Registrar General of India, Government of India, New Delhi

Robert, Fioramonti, Kubiszewski I (2016) The UN sustainable development goals and the dynamics of well-being. Front Ecol Environ 59. Available on: https://doi.org/10.1002/fee.1231

Roberts K (1980) Price-independent welfare prescriptions. J Public Econ 13:277–297. Available on: https://www.sciencedirect.com/science/article/abs/pii/0047272786900071

Samuelson PA, Swamy S (1974) Invariant economic index numbers and canonical duality: survey and synthesis. Am Econ Rev 566–593. Available on: http://www.jstor.org/stable/1813311

Scherer L, Behrens P, Koning A, Heijungs R, Sprecher B, Tukker A (2018) Trade-offs between social and environmental sustainable development goals. Environ Sci Policy 65–72. Available on: https://doi.org/10.1016/j.envsci.2018.10.002

SECC (2011) SECC 2011 at a glance. Ministry of Rural Development, Government of India, New Delhi. Available on: https://secc.gov.in/getSeccDataSummaryNationalReport.htm

Sen A (1976) Poverty: an ordinal approach to measurement. Econometrica 219–231

Sen A (1979) Equality of what? The Tanner lecture on human development, 22 May, pp 195–220

Sen A (1999a) Development as freedom. Knopf, New York

Sen A (1999b) Development as freedom: an India perspective. Indian J Ind Relat 157–169

Sen A (2001) Many faces of gender inequality. Frontline. Available on: https://www.sas.upenn.edu/~dludden/MANY%20FACES%20OF%20GENDER%20INEQUALITY.htm

Sen A (2005) The three R's of reform. Econ Polit Wkly 1971–1974. Available on: http://www.jstor.org/stable/4416610

Sen A, Dreze J (2011) Putting growth in its place: it has to be but a means to development, not an end in itself. Outlook, 14 Nov. Available on: https://www.outlookindia.com/magazine/story/putting-growth-in-its-place/278843

Shah M (2007) Employment guarantee, civil society and Indian democracy. Econ Polit Wkly 43–51. Available on: http://www.jstor.org/stable/40276832

Sharma R (2018) Health and economic growth: evidence from dynamic panel data of 143 years. PLoS ONE. https://doi.org/10.1371/journal.pone.0204940

Sharma AN, Saluja MR, Sharma A (2016) Macro economic impact of social protection progrrames in India. Econ Polit Wkly 121–126

Singh M (2012) PM's opening remarks at the meeting of PM's national council on skill development. Prime Minister Office, New Delhi. Available on: https://archivepmo.nic.in/drmanmohansingh/content_print.php?nodeid=1126&nodetype=2

Singh UK (2021) India has a serious hunger problem and it needs urgent policy intervention. The WIRE, 30 Oct. Available on: https://thewire.in/food/india-has-a-serious-hunger-problem-and-it-needs-urgent-policy-intervention

Singh NK, Stern N (2013) The new Bihar. HarperCollins

Smith P, Olesen JE (2010) Synergies between the mitigation of, and adaptation to, climate change in agriculture. Agric Sci 543–552. Available on: https://doi.org/10.1017/S0021859610000341

TARA (2015) Achieving the sustainable development goals in India: a study of financial requirements and gaps. Technology and Action for Rural Advancement, New Delhi. Retrieved from: https://www.devalt.org/images/L3_ProjectPdfs/AchievingSDGsinIndia_DA_21Sept.pdf

Tewary A (2020) Tejashwi Yadav's apology for 'mistakes' of RJD regime fails to impress Bihar's ruling NDA. The Hindu, 3 July. Available on: https://www.thehindu.com/news/national/other-states/tejashwis-apology-for-mistakes-of-rjd-regime-fails-to-impress-bihars-ruling-nda/article31982878.ece

Thapar R (1966) A history of India. Penguin Books, London

Thomas D, Frankenberg E (2002) Health, nutrition and prosperity: a microeconomic perspective. Bull World Health Organ 106–113

UN (1972a) Report of the United Nations conference on the human environment. United Nations, New York. Available on: https://undocs.org/en/A/CONF.48/14/Rev.1

UN (1972b) United Nations conference on the human environment, Stockholm, 5–16 June 1972. United Nations, New York

UN (1987) Report of the world commission on environment and development: our common future. United Nations, New York. Available on: https://sustainabledevelopment.un.org/content/documents/5987our-common-future.pdf

UN (1992a) Report of the United Nations conference on environmemt and development, vol I. United Nations, New York. Available on: https://undocs.org/en/A/CONF.151/26/Rev.1

UN (1992b) United Nations conference on environment & development: agenda 21. United Nations, New York

UN (1992c) United Nations conference on environment and development. United Nations, New York

UN (2002) Report of the world summit on sustainable development. United Nations, New York. Available on: https://undocs.org/pdf?symbol=en/A/Conf.199/20

UN (2012a) Resolution adopted by the General Assembly on 27 July 2012. United Nations, New York. Available on: https://www.un.org/ga/search/view_doc.asp?symbol=A/RES/66/288&Lang=E

UN (2012b) Resolution adopted by the General Assembly on 27 July 2012. Agenda No. A/RES/66/288. United Nations, New York

UN (2013) Open working group of the General Assembly on sustainable development goals. Agenda No. A/67/L.48/Rev.1. United Nations, New York

UN (2015a) Resolution adopted by the General Assembly on 25 September 2015. A/RES/70/1. United Nations, New York

UN (2015b) Resolution adopted by the General Assembly on 25 September 2015. Agenda No. A/70/L.1. United Nations, New York. Available on: https://www.un.org/ga/search/view_doc.asp?symbol=A/RES/70/1&Lang=E
UN (2015c) When we stand together, there is no limit to what we can achieve, Ban tells UN Assembly. United Nations, New York. https://www.un.org/sustainabledevelopment/blog/tag/secretary-general-ban-ki-moon/
UN (2016a) MDG monitor: MDG 6: combat HIV/AIDS, malaria and other major diseases. United Nations, New York
UN (2016b) Report of the inter-agency and expert group on sustainable development goal indicators. Economic and Social Council, United Nations, New York. Available on: https://undocs.org/E/CN.3/2016/2/Rev.1
UN (2017) The sustainable development goals report: 2017. United Nations, New York. Available on: https://unstats.un.org/sdgs/files/report/2017/thesustainabledevelopmentgoalsreport2017.pdf
UN (2020a) Policy brief: the impact of COVID 19 on children. United Nations, New York. Available on: https://unsdg.un.org/sites/default/files/2020-04/160420_Covid_Children_Policy_Brief.pdf
UN (2020b) The sustainable development goals report: 2019. United Nations, New York. Available on: https://unstats.un.org/sdgs/report/2019/The-Sustainable-Development-Goals-Report-2019.pdf
UN (2020c) United Nations in India. United Nations, New York. Available on: https://in.one.un.org/page/sustainable-development-goals/gender-equality-sdg-5/
UN (2021) The sustainable development goals report: 2021. United Nations, New York. Available on: https://unstats.un.org/sdgs/report/2021/
UNAIDS (2021) Global HIV & AIDS statistics—fact sheet. Retrieved from: https://www.unaids.org/en/resources/fact-sheet
UNDP (1997) Human development report 1997. United Nations Development Programme, New York
UNDP (2015) Building the post-2015 development agenda: 2014 annual report. United Nations Development Programme, New York. Available on: https://www.undp.org/sites/g/files/zskgke326/files/publications/UNDP_Building%20the%20Post-2015%20Development%20Agenda_Final_2015.pdf
UNDP (2018) Human development report 2018. United Nations Development Programme and Oxford University Press, New York
UNDP (2019) Global multidimensional poverty index 2019: illuminating inequalities. United Nations Development Programme and Oxford Poverty and Human Development Initiative, New York. Available on: http://hdr.undp.org/sites/default/files/mpi_2019_publication.pdf
UNICEF (2016) A snapshot of Bihar elementary education: district-wise statistical report. UNICEF BIHAR, Patna. Available on: https://www.bepcssa.in/en/A%20Snapshot%20of%20Bihar%20Elementary%20Education1.pdf
UNICEF (2017) A world free from child poverty: a guide to the tasks to achieve the vision. United Nations Children's Fund, New York
UNICEF (2020a) Maternal mortality in 2000: estimate developed by UNICEF and UNFPA. United Nations Children's Fund, New York
UNICEF (2020b) Universal child benefits: policy issues and options. UNICEF, New York. Available on: https://www.unicef.org/media/72916/file/UCB-ODI-UNICEF-Report-2020.pdf
UNICEF India (2016) Children in India: India's growth over the last two decades has contributed phenomenally to global human development. United Nations Children's Fund, New Delhi
Wheebox (2021) India skills report 2021. Wheebox, Gurugram. Available on: https://indiaeducationforum.org/pdf/ISR-2021.pdf
WHO (1990) Maternal mortality in 2000: estimate developed by UNICEF and UNFPA. World Health Organization, Geneva
WHO (1991) Maternal mortality: a global factbook. World Health Organization, Geneva
WHO (2019) Maternal mortality. World Health Organization, Geneva
WHO (2020) Global report on WASH in health care facilities. World Health Organization, Geneva

World Bank (1993) World Bank development report 1993: investing in health. The World Bank Group & Oxford University Press, Washington, DC

World Bank (2008) World development report 2008: agriculture for development. The World Bank, Washington, DC. Available on: https://doi.org/10.1596/978-0-8213-6807-7

World Bank (2009) Doing business in India 2009. The World Bank, Washington, DC. Available on: https://openknowledge.worldbank.org/handle/10986/28088#:~:text=Doing%20business%20in%20India%202009%20is%20the%20first%20country%20specific,cities%2C%20and%20documents%20their%20progress

World Bank (2017) New World Bank project to give impetus to India's skill's agenda—to provide over 8 mn youth with market-relevant training. The World Bank Group, Washington, DC

World Bank (2020) Global economic prospects, June 2020. The World Bank, Washington, DC. Available on: https://openknowledge.worldbank.org/handle/10986/33748

World Bank (2021) World Bank indicators. The World Bank, Washington, DC. Available on: https://data.worldbank.org/indicator/SP.DYN.IMRT.IN

Worldometers (2020) Coronavirus cases. Worldometers. Available on: https://www.worldometers.info/coronavirus/coronavirus-cases/?fbclid=IwAR1rXCB1NYf48HuNqhHZDPNSUySwz8PpWwqaGRsE-3PijhQfc03axd2FPcI

Zhao Z, Cai M, Wang F, Winkler JA, Connor T, Chung MG, Zhang J, Yang H, Xu Z, Tang Y, Ouyang Z, Zhang H, Liu J (2021) Synergies and tradeoffs among sustainable development goals across boundaries in a metacoupled world. Sci Total Environ 1–18. Available on: https://doi.org/10.1016/j.scitotenv.2020.141749

Chapter 6
Impact of COVID-19 on Sustainable Development Goals

Abstract COVID-19 pandemic brought the whole world to a standstill. As governments all over the world struggle to grapple with its debilitating impact, they have expended all resources at their disposal to mitigate the adverse impact of the pandemic on human development. In the wake of this, the goals set upon by the Sustainable Development Goals are in danger of taking a backseat. In view of this reality, this chapter emphasizes the need to incorporate the noble SDG targets during the preparation of COVID-19 strategies.

Keywords India · COVID-19 · SDGs · Poverty · Health

6.1 Introduction

Humankind is pursuing sustainable development goals (SDGs) 2030, as the member countries of the United Nations Organization have, from 2016, embarked on their journeys to finish the unfinished targets of Millennium Development Goals (MDGs). Agenda 2030 is to leave no one behind by the target year of 2030. The 17 SDGs and 169 targets are based on 5 elements: people, planet, peace, prosperity, and partnership. Thus translates into a world where ever body has adequate food to eat, proper facility to access advanced medical facility, quality education, decent job, and better environment. The spirit of partnership between member countries remains the bedrock for achieving all these goals and building a new world by 2030.

The prime focus of SDGs is on poverty eradication (SDG 1). The challenges for poverty reduction are particularly severe in developing and underdeveloped economies. Most of these countries fall in Africa, Asia, and Latin America. Asia is the home to 61% of the global population while Africa constitutes 17%. United Nations Development Program (UNDP) use Multi-Dimensional Poverty Index to measure the progress against SDG 1, which aims to end poverty in all its forms. HDR Report (2019) highlights that among the following 10 countries (Bangladesh, Cambodia, India, Niger, Pakistan, Vietnam, Democratic Republic of Congo, and Haiti) belonging to different income levels and comprising 2 billion population finds that the fastest absolute reduction in multidimensional poverty is visible in India, Cambodia, and Bangladesh, followed by Ethiopia and Haiti.

U. K. Singh, *Human Development in Bihar, India*, Advances in Geographical and Environmental Sciences, https://doi.org/10.1007/978-981-19-3624-1_6

In between, however in 2020, the COVID-19 pandemic has put the world economy on standby. What does this mean for the developing world and their quest for attaining sustainable development remains the core concern of this paper. In the developing world, two different countries exist in one. As the corona virus spread rapidly in the first half of 2020, the emerging economies of Asia and Africa resorted to imposing stringent lockdown/shutdown as a measure to mitigate the spread of the virus as well as to buy some valuable time in beefing up the state capacity. Lockdown has had differentiated and even contrasting meanings for different classes. For the wealthier people, living in urban and suburban areas, it means no jogging specifically no dog-walking but for the relatively poorer sections of the society, it means no job to survive since most of the people are employed by/in the informal sector of the economy. Policing in the developing world too is harsh as the disadvantaged may often have to taste the sticks in order to go out even for buying essential commodities. The police force is clearing roads and maintaining social distancing norms, which is simply not possible in densely populated areas. For instance, India imposed a lockdown to ensure that movement is not restricted for essential commodities. As reported in STATICA, 1380 million people in India are placed in enforced lockdown due to the pandemic. Unregulated settlements of these countries were worse affected as they lack access to clean drinking water and proper sanitation facilities. The government is scared because of the fear of community spread in densely populated areas. If the virus spread uncontrollably it is almost certain that government will have no vacant beds in hospitals and the human cost would be enormous in that situation. The further looming economic crisis is clearly visible. Especially for the million depending on the informal sector.

In India, 400 million people are employed in the informal sector and most of them are now on verge of falling into poverty because of pandemic-induced crisis. India at present is struggling hard to take its people out of poverty, curb the menace of unemployment, make them food secure, equip them with quality education, provide decent jobs and decrease the dependency on fossil fuels. This chapter will discuss the impact of the COVID-19 pandemic crisis on SDGs in India.

6.2 Poverty

Poverty is a situation when there is no food to eat; source of income is dried, house has run out of utensils, no proper clothing to wear, and even no roof over head. Poverty can also be realized even when the belly is full, but not with nutrient-rich food. For many years, poverty measurement has attracted scholars of development studies. Initially, poverty measurement broadly includes the following two steps: identifying the number of poor out of the total population and constructing an index on available data. Identification based on the income concept centers on whether an individual touches the minimum threshold line is defined as the poverty line (Sen 1976). Alkire and Roche (2013) presented a practical approach to poverty measurement and emphasized its multidimensional nature, thus requiring qualitative data (literacy and physical security), along with continuous and cardinal data (income and consumption). Multidimensional approach is, therefore, made up of three dimensions, named as the three dimensions of poverty (Health, Education, and Standard of Living) with ten indicators (Nutrition, Child Mortality, Years of Schooling, School attendance, Cooking fuels, Sanitation and Drinking Water, electricity, Housing, and Asset). Every dimension of multidimensional poverty is equally weighted and each indicator within a dimension is also equally weighted. If an individual is deprived in at least one-third of indicators he/she will be considered poor (UNDP 2018). As SDGs are interconnected so we can say that SDG 17 (Partnership for development) is all about learning from the expertise of each other—enabling partners to make use of the latest and the best techniques to achieve their goals. Hence, India which shares an almost equal percentage of the world population with Africa can be said to have a special onus of leading by example and also taking Africa's population out of poverty with its own expertise.

Sudden economic deterioration in advanced economies and major emerging markets and developing economies have had a massive impact on export and other industries of the South Asian Region, while the nationwide shutdown has led to a sharp contraction in domestic consumption. Though the pandemic reached South Asian Region relatively delayed, the rate of the spread of infection has been much higher and cases are still rising rapidly. Collapsed global demand has plunged the economic activity of the region as export-oriented industries are in disarray. This is reflected in sharp decelerations in the purchasing managers' indexes and new export orders in India, the largest regional power. Trade has sharply fallen across the world. Sales and production in a number of key sectors in regional economies have been hit especially hard amid anemic demand. Business confidence in both manufacturing and services sectors has concomitantly fallen. Key trading corridors in the region have also witnessed disruptions. Private consumption has been severely hindered in Bangladesh, India, Nepal, and Pakistan. Closure of small and medium-sized enterprises, a key engine of regional private sector activity, induced substantial loss in employment and private investment. Some recent relaxations to these measures have been cautious, given the continued rise in COVID-19 cases (World Bank 2020).

Till 18th June 2020, the number of confirmed COVID-19 cases has risen to around 1,601,949 with more than 53,669 people having died in the South Asian Region (Worldometers 2020). While limited testing capacity may understate the true scale of the regional outbreak, the majority of infections in the region are in India (1,077,864), Pakistan (263,796), and Bangladesh (202,066) with the reported number of deaths being 26,828, 5568, and 2581, respectively. Complete nationwide shutdown in these three largest regional economies sharply curtailed activity in the services sector and manufacturing production.

World Bank has estimated that the C0VID-19 pandemic will push around 60 million people into extreme poverty, globally. World Bank President, David Malpass has said that the pandemic and lockdowns will erase all the advancements done to poverty eradication and could put as much as 60 million people into extreme poverty in advanced economies in 2020 increasing the number of the world's population living on less than $1.90 per day toward 700 million people and causing the global poverty rate to rise for the first time since 1998, when the Asian Financial Crisis hit (World Bank 2020).

Sustainable Development Goals inherit these dimensions in various goals, for instance, SDG 1 is about poverty, SDG 2 talks about nutrient food, SDG 3 is about ensuring good health, SDG 4 pledges to provide quality education, SDG 6 to promote access to safe drinking water and sanitation, SDG 7 seeks to increase access affordable clean energy, and SDG 11 to provide access to safe housing.

6.3 Impact of Pandemic on Goals

SDG 1 talks about poverty reduction. India, along with China leaves a clear imprint on poverty reduction. UNDP (2019) applauded the role of India's pro-poor welfare scheme during the period of millennium development goals in accelerating the reduction of poverty among them ranked as 101. India remarkably added assets, improved cooking fuel, sanitation facility, and nutrition. A unique feature of this improvement has been that India has done better in rural areas as compared to urban areas. The share of people with multidimensional poverty has come down to 27.9% in 2015–16 from 55.1% in 1998–99.

Pandemic attracted attention to the potential magnitude of shrinking of economic activity in the country. As a consequence of the lockdown, there has been a massive impact on employment and labor. Developing countries have received less attention. India is a middle-income developing country where millions of people live just above the poverty line. A recent research conducted by Kings College London and Australian National University found that poverty is increasing dramatically in middle-income developing countries. Further, the report added that with expanding poverty line additional poor were concentrated in those areas where the corresponding poor under 20% concentration scenario. Middle East and North Africa along with Latin America will shelter 18%. While India played a significant role in driving down the potential increase in global extreme poverty documented previously, it still comprises almost half the estimated additional poor regardless of the contraction scenario.

World Bank growth estimation is presenting a gloomy picture of the Indian economy. As per estimates, its growth had slowed down to 4.2% in FY-20, which ended in March 2020. Output is projected to contract by 3.2% in FY 2020/21, when the impact of the pandemic will largely span out. Stringent measures to control the spread of the virus will heavily curtail activity, despite some support from fiscal and monetary stimulus. Spillovers from weaker global growth and balance sheet stress in the financial sector will also weigh on activity (World Bank 2020).

SDG 2 lays emphasis on the importance of food security to achieve zero hunger, and is linked to the achievement of **SDG 1**, which has been discussed at length above. Hunger is caused by poverty and inequality, not scarcity. For the past two decades, the rate of global food production has increased faster than the rate of global population growth. The world already produces more than 1½ times the food to feed everyone on the planet. That's enough to feed 10 billion people, the population peak expected to be achieved by 2050. But the people making less than $2 a day, most of whom are resource-poor farmers cultivating at unviable small plots of land who can't afford to buy this food. President of the International Fund for Agriculture (IFAD), pointed out that "there are 7.8 billion people in the world. We grow enough food to feed 10 billion", thus hinting at the issues of "mismanagement" of food and affordability/access as constituting the prime sources of current hindrances to achieving SDG 2.

A recent report titled "The State of Food Security and Nutrition in the World" presents a grim picture with respect to achieving SDG 2 amidst the COVID-19 pandemic (FAO 2020). The report is a combined work of various agencies (FAO, WHO, IFAD, UNICEF, and WFP) involved with the United Nations and have cautioned against the world being "off track" of achieving the goals of ending hunger, food insecurity, and all forms of malnutrition. Quite predictably, the coronavirus pandemic has made things worse by further "intensifying the vulnerabilities and inadequacies of global food systems", the report has noted. WHO Director-General Adhanom Ghebreyesus has said that "While it is too soon to assess the full impact of COVID-19, the report estimates that 130 million more people may face chronic hunger by the end of this year". Asia remains home to the largest quantity of undernourished people (381 million) in the world, followed by Africa (250 million) and Latin America and the Caribbean (48 million) (FAO 2020).

In the context of India, while the country has achieved tremendous economic growth but still hunger and malnutrition exist. Food security continues to remain a major challenge as out of the 1.3 billion population 195 million or 14.8% of the Indian population remain undernourished, making India home to the largest number of undernourished people in the world (FAO, IFAD, UNICEF, WFP and WHO 2018).

Despite these facts, the first and foremost thing that must be highlighted is that India is not facing a situation of food scarcity. Quite contrary to that, after the success of the Green Revolution, the country has made tremendous strides in improving the overall production and has not only attained self-sufficiency in the case of food grains but is also a major exporter of food items. India has become a food exporter from being a food importer. The government warehouses currently remain stacked, rather oversupplied, to the tune of three times more than the buffer stock norms of 21.04 million metric tons, with grains. This was right before the onset of the sowing of Kharif crops. Yet, as various statistics prove, hunger, malnutrition, and poor health outcomes remain perennial afflictions plaguing Indian society. Global Hunger Index (2019) report depicts that 38.4% of children aged under five in India are stunted (too short for their age), while 21% suffer from wasting, meaning their weight is too low for their height.

It is in this respect that the COVID-19 crisis is going to hit the Indian masses the hardest. As reports of people, mainly migrant laborers enroute to their native places, dying across the country due to hunger have stormed the print and broadcast platforms, the "invisible" problem of unequal access to food has gained "visibility".

Such stories and images have even gained traction with global media giants like the New York Times, Washington Post, etc.

Ensuring food security is about concomitantly achieving three objectives: food availability, food access, and food utilization or absorption. COVID-19 crisis will adversely affect all three. Food availability will be impacted due to labor shortages and disruption of supply chains of necessary inputs like seeds, fertilizers, etc. However, this is the least difficult of all the challenges as with even limited government support and technological interventions, this problem could be surmounted at least in the short term. Of course, if the crisis persists for a longer duration then this most foundational or basic issue can also become a critical problem. The real problem, however, lies in tackling the second and third factors viz., ensuring access/affordability when job loss is rampant, and food absorption/utilization. Food absorption, as health experts throughout the world have opined, assumes great importance because the rate of fatality due to coronavirus is directly linked to weak immunity, which in turn is a direct consequence of not having access to quality and nutritious food. In that regard, India seems to be placed in a rather precarious situation. According to health experts, proteins constitute the most critical nutrients that support our immune system but since a vast majority of Indians are vegetarians so they generally tend to consume lesser amounts of protein. Besides, the fact that a protein diet is costlier makes it an outrightly unaffordable option for a vast chunk of people. This then calls for the government to not only raise awareness about the necessity of increasing the protein intake but also enlarge the PDS basket to include food items rich in protein. It has been a welcome step to include pulses in the PDS pool.

SDG 3 talks about healthy life and promoting overall good health, with **SDG 3.4** talking about the promotion of mental health and well-being. With easing norms of lockdown, India is heading toward a new epidemic "mental illness". The Indian Psychiatry Society, 20% more cases were reported in the initial week of lockdown (Loiwal 2020). In the coming weeks and months, the country is slated to witness a huge surge in the number of mental health issues, especially because of rising unemployment, alcohol abuse, economic hardship, domestic violence, and indebtedness. The impact will be more severe on poor, most vulnerable, and marginalized section of society though its impact is observed on the entire population. Mental health has been another area whose criticality has been exposed in the COVID-19 pandemic for which India is not prepared. India's mental health system is not well equipped as only 9000 specialists are available for the 1.3 billion population. Pandemic exposed the dilapidated situation of mental health infrastructure and shows the gap between demand and supply for the same. The looming threat of sudden death and the associated fear of psychosis rampant among the masses due to a complete lack of understanding of the virus especially in the initial days created an atmosphere of uncertainty. This uncertainty has been the single biggest source of anxiety and psychological trauma for individuals. The importance of what Aristotle said in the ancient past about human nature when he wrote that "Man is by nature a social animal" has been put to test during the global pandemic. While technological solutions have offered a way to remain "connected" with the world out there but its limitations too have been made apparent. The virtual can complement the real but it cannot be a substitute for the latter. Also, lockdown, in a world characterized by unequal power relations, especially within the household, has meant that its effects have been unequal, contrasting, and even discriminatory, for different people. For instance, a UK-based study has shown that the cases of domestic violence have increased rampantly during the lockdown period. Taking a cue from that one can safely extrapolate that the situation would be somewhat similar in other areas as well, including India.

On the brighter side, however, one can say that this has also been a period of introspection for individuals and societies collectively. Though, only time will tell if humankind has learned its lessons or not. It's an old adage that we realize the importance of something only when we lose it. This lockdown has shown the human species the preponderance of freedom to move freely and its centrality to mental well-being, which they have hitherto taken for granted. This realization must not be forgotten especially in the post-corona world.

SDG 4 is about quality education. Nelson Mandela once said that education is the only weapon to uplift the people from deprivation. The complete shutdown imposed by 188 countries pushed the vulnerable and marginalized population into extreme poverty, the number of people who fell into extreme poverty is nearly 42–65 million. Lockdown affected the education system at length, almost 1.5 billion youth and children were deprived of education (UN 2020a, b, c).

Online teaching and learning emerge as an option for physical classes. The digital divide between rich and poor in developing countries failed the online teaching platform as a solution at a time when people are not having access to internet services. International Telecommunication Union (ITU) reported that in India, 20.1% of individuals use internet facilities from any location, further data for accessibility of internet services by gender is starker, only 14.9% access internet facility against 25% of their male counterpart (ITU 2021). The digital divide among gender translates into the deprivation of girls from education and negatively affect global goals.

Education Ministry, Government of India, has found that 14,336,007 children of Bihar are not having any type of digital device, which constrained the opportunity of primary attaining education (Ministry of Education 2021).

Literacy is basic means of survival in fast-changing world. It's a cushion for further study. Chances of exploitation increase with illiteracy; children falling into economic misery translates into lifetime weakness. The occurrence of mass illiteracy, contained the progress of human development (Dreze 2021).

At the onset of the pandemic, the concern for graduating students is related to their placement, prolonged lockdown limited their prospect of getting hired or delayed in joining. India Skill Report (2021) shows that the proportion of employment among graduating students in India is declined in 2021 from 2020; the shock of stringent containment measures resonates in the proportion of job cuts across all the leading sectors such as automotive (−54), core and energy (−45), engineering and manufacturing (−36), others and diversified (−62).

Pandemic has transformed student–teacher relation, by taking the learning process "online". The situation pushed the policy-makers across the globe to find a solution to the digital divide, by ensuring inclusive e-learning for all. Digital Empowerment Foundation figured that 30% of India's population are illiterate while triple the rate for digitally literate. To deliver digital literacy, in 2015, the Government of India had launched Digital India on mission mode. Although, good intentions of the government flop on the ground, because of delayed infrastructure development, availability of bandwidth, volume of personal computer, and the scale of capacity. As per information shared by the Telecom Regulatory Authority of India (TRAI), the number of wireless subscribers in India is 115 crore, out of which 66 crore subscribers access to quality bandwidth. If we consider the broadband speed equal to or above 512 kbps, half of the Indians are deprived of access to good quality internet bandwidth. So it is clear that in India, quality internet access is for the privileged class and not for everyone.

SDG 5 is about gender equality, early empirical evidence show that women may be biologically better equipped than men to fight against infections, notwithstanding the socioeconomic consequences of COVID-19 against women. This is a common scenario in developing countries like India, where gender discrimination persistent challenge. Pandemic is changing domestic dynamics. Household responsibility of care economy like cooking, cleaning, rearing child have expanded in the wake of the current crisis. Another detrimental feature of eating last is that they (women) have the least nutrient intake because of financial constraints and food shortages. The same pattern is visible across the developing world, from Zimbabwe to Bangladesh.

Lockdown has locked all institutional and organized and unorganized support for women. In the developing world, low-income households share a single smartphone owned by a male counterpart, making them more vulnerable to misinformation.

Digital divide will fuel other consequences for women, especially in education and employment. With a limited data pack, girls in countries like India may lose out on online education, given that they are less likely than boys to have access to the internet. And as analysts foresee more jobs moving online post-pandemic, the digital divide might exacerbate job market inequalities.

Surely, in the post-COVID-19 world, economic fallouts are likely to reduce job prospects for women.

With ease in lockdown, businesses are allowed to operate with fewer employees, and trends toward mechanization could also accelerate. Since women are generally consigned to menial tasks within production processes, with the automation of firms women may be the first to lose jobs. And with partial lockdowns involving reduced public transit, women might find their access to work curtailed, given their disproportionate reliance on buses and trains for commuting.

Most of the state governments have made horizontal reservation arrangements of 33% to increase the number of women in non-farming activities. Bihar government had taken a progressive step, reserving 35% of seats at all levels in government jobs through direct recruitments (mint 2016).

Gender parity in employment is a bigger concern of policy-makers of India. In India, most of the women (15–59 age group) are engaged in domestic works, an estimated 61.6% in rural areas and 65.3% in urban areas are employed as a domestic worker; the proportion of women in domestic work is the highest in Bihar, the rural–urban proportion are 82 and 74; the proportion of women population ratio in during 2019–2020 was 19% and 14.5% in urban and rural areas, respectively; Maximum number of women are engaged in transport & communication (45.6%), manufacturing trade (24.5) and hospitality (13.8%) (NSSO 2014, 2020). If female labor participation continues to decline in post-COVID phase, it will damage the stature of women not only in their families but will also hamper the national economic growth, as the full participation of half population is required to achieve tall development goals.

SDG 6 seeks to ensure the availability of proper sanitation facilities and clean drinking water. The African proverb says that the disease of a poor man takes him to his grave. This is a very real danger in times of pandemic when health infrastructure is not sufficient and does not cater to a large section of the population. In tough times prevention becomes better than cure.

In tough times, prevention become better than cure. To contain the spread of the COVID-19 virus, proper hygiene is important. The virus transmits through any physical contact; virus travels with infected person from one place to another. In order to reduce transmission of COVID-19 virus, frequent hand wash is needed, as suggested by medical practitioners across the world. The practice also helps in breaking the transmission chain of other vector-borne diseases (WHO 2020).

Around two decades ago the then Vice President of the World Bank, Ismail Serageldin, had opined that "the next world war will be over water". This simple yet profound statement had since then become a clichéd line used in all discussions surrounding conflicts over water within and among nations. The centrality of water to human survival on the planet is so foundational that it need not be explicitly mentioned, but still, one in three people globally (in 2019) did not have access to safe drinking water as per UNICEF, and WHO. Therefore, the Sustainable development Goal (SDG) 6 aims to provide access to safe drinking water and sanitation to all by 2030. In India where owing to a large population, the per capita pressure on natural resources is even more acute, the situation is truly horrific. A NITI Aayog report released in 2018 had predicted Day Zero for 21 Indian cities by 2020. Day Zero refers to the day when a place is likely to have no drinking water of its own. Similarly, according to the Composite Water Management Index of the NITI Aayog, 75% of households do not have drinking water on-premise and about 84% of rural households do not have piped water access. It is in this situation that the COVID-19 crisis has hit the country. WHO has been frequently giving out guidelines highlighting the necessity of ensuring proper hygiene, especially constant hand-washing, as measures to tackle the spread of coronavirus. But, given the overall lack of availability as well as wide regional imbalances in the availability of water resources has compounded the problem for common people. On top of that, especially in poorer states like Bihar, Jharkhand, etc., a large chunk of ultra-poor does not have the purchasing power to even buy soaps or sanitizers. Thus, for the majority, COVID-19 has come as a double whammy as they do not even have the "luxury" to avail things required to ward off the grave threat.

As SDGs are interconnected and impact on SDG 8, SDG 9, SDG 10 SDG 11, SDG 12 has already been discussed above, SDG 13, SDG 14, SDG 15 (environment related) are likely to be "positively" impacted, however only in the short term, due to lockdowns, as they have restricted human movement.

SDG 17 is about partnership for development. Abrahamsen (2004) argued that partnerships leverage technology transfer in a global liberal framework to sustain the governance of states and individuals in a liberal democratic framework. The word partnership shows that the donor no more want to impose their vision, they want partnership for development. SDG 17 provides a platform for the member countries of the UN to forge a partnership for human development in the frame of global goals. During the test time of the COVID-19 pandemic, the international community, NGOs, and private entities come together to support the global cause.

6.4 Conclusion

Meanwhile, as far as the achievement of these goals by India is concerned then as the aforementioned work has shown, it will all be going to be an uphill task from here onwards. The need of the hour is to boost local capacity as those working on the ground remain the best line of defense against the unfolding catastrophe. Governments at all three levels must work in collaboration to beef up the social infrastructure in general and of the health sector in particular. Issue of livelihood preservation, poverty reduction, gender equality, good health, and sanitation all remain intricately connected and so positive interventions in one area are bound to have ripple effects in other domains. The government in partnership with the people must take all the necessary steps, some of which have been alluded to in the above discussion, in mitigating the current disaster.

References

Abrahamsen R (2004) The power of partnerships in global governance. Third World Q 1453–1467

Adhikari A, Bhatia K (2010) Can we bank on banks? Econ Polit Wkly

Afridi F, Mukhopadhyay A, Sahoo S (2016) Female labour force participation and child education in India: the evidemce from national rural employment gurantee scheme. IZA J Lab Dev 1–27

Ahluwalia MS (2020) Economic performance of states in post-reforms period. Econ Polit Wkly 1637–1648

Alkire S, Roche JM (2013) How multidimensional poverty went down: dynamics and comparisons. University of Oxford, Oxford. Available on: https://www.ophi.org.uk/wp-content/uploads/How-Multidimensional-Poverty-went-down.pdf

Allen C, Metternicht G, Wiedmann T (2019) Prioritising SDG targets: assessing baselines, gaps and interlinkages. Sustain Sci 14:421–438. Available on: https://link.springer.com/article/10.1007%2Fs11625-018-0596-8

Anand R, Tulin V, Kumar N (2014) India: defining and explaining inclusive growth and poverty reduction. IMF working paper

Bank W (2019) The state of social safety nets: 2018. The World Bank Group, Washington, DC
Benassy-Quere A, Bourguignon F, Dercon S, Estache A, Gunning JW, Kanbur R et al (2010) The millennium development goals: an assessment. In: Spence RK (ed) Equity and growth in a globalizing world. The World Bank, Washington, DC, pp 17–40. Available on: https://openknowledge.worldbank.org/bitstream/handle/10986/2458/548910PUB0EPI11C10Dislosed061312010.pdf?sequence=1&isAllowed=y
Bhalla SS (2013) The unimportance of NREGA. Indian Express, July 24. Available on: https://indianexpress.com/article/opinion/columns/the-unimportance-of-nrega/
Bloom DE, Canning D, Jamison DT (2004) Helath, wealth and wellfare. Financ Dev 10–15
Bue MC, Klasen S (2013) Identifying synergies and complementarities between MDGs: results from cluster analysis. Soc Indic Res 647–670. Availble on: https://link.springer.com/article/10.1007%2Fs11205-013-0294-y
Census (2011) Census of India 2011. Registrar General & Census Commissioner, India, New Delhi. Available on: https://censusindia.gov.in/2011Census/pes/Pesreport.pdf
Chakraborti D, Rahman MM, Chatterjee A, Das D, Das B, Nayak B, Roy MM (2016) Fate of over 480 million inhabitants living in arsenic and fluoride endemic Indian districts: magnitude, health, socio-economic effects and mitigation approaches. J Trace Elem Med Biol Organ Soc Miner Trace Elem (GMS) 33–45. Available on: https://doi.org/10.1016/j.jtemb.2016.05.001
Deaton A (2003) Health, inequality, and economic development. J Econ Lit 113–158
Desai S, Vashishtha PS, Joshi OS (2015) Mahatma Gandhi national rural employment guarantee act: a catalyst for rural transformation. National Council of Applied Economic Research, New Delhi
Desjardins R, Schuller T, Feinstein L (2006) Understanding the social outcomes of learning. In: Desjardins R, Schuller T, Feinstein L (eds) Measuring the effects of education on health and civic/social engagement. OECD, pp 11–15
Dreze J (2021) Where is the strategy for dealing with learning loss during Covid? The Indian Express, 17 Sept. Available on: https://indianexpress.com/article/opinion/columns/where-is-the-strategy-for-dealing-with-learning-loss-during-covid-7511344/
Dreze J, Dev N, Khera R (2008) Corruption in NREGA: myth and reality. The Hindu, Jan 22
Drèze J, Gupta P, Khera R, Pimenta I (2019) India's public distribution system after the food security act. Econ Polit Wkly 36–47
Engström RE, Destouni G, Howells M, Ramaswamy V, Rogner H, Bazilian M (2019) Cross-scale water and land impacts of local climate and energy policy—a local Swedish analysis of selected SDG interactions. Sustainability 242–247. https://doi.org/10.3390/su11071847
Esteves T, Rao K, Sinha B, Roy SS, Rao B, Jha S et al (2013) Agricultural and livelihood vulnerability reduction through the MGNREGA. Econ Polit Wkly 94–103. Available on: http://www.jstor.org/stable/24477901
FAO (2020) The state of food security and nutrition in the world. Food and Agriculture Organization, Rome
FAO, IFAD, UNICEF, WFP, and WHO (2018) The state of food security and nutrition in the world: Building climate resilience for food security and nutrition. Rome: FAO
Ferreira FH, Lugo MA (2013) Multidimensional poverty analysis: looking for a middle ground. World Bank Res Obs 220–235. Available on: https://doi.org/10.1093/wbro/lks013
Fukuda-Parr S (2004) Millennium development goals: why they matter. Glob Gov 10(4):395–402
Global Hunger Index Report 2019 (2019) Cocern Worldwide, Dublin
Government of Bihar (2011) The Bihar state free and compulsory education of children rules, 2011. Human Resource Development Department, Government of Bihar, Patna. Available on: https://www.education.gov.in/sites/upload_files/mhrd/files/upload_document/bihar-rte-rules_english%20(1).pdf
Government of Bihar (2016) Economic survey 2015–16 of Government of Bihar. Ministry of Finance, Governmnet of Bihar, Patna. Available on: https://www.adriindia.org/centre/report_details/economic-survey-2015-16-of-government-of-bihar

Haines A, Smith KR, Anderson D, Epstein PR, McMichael AJ, Roberts I et al (2007) Policies for accelerating access to clean energy, improving. The Lancet 370:1264–1281

Hazarika I (2012) India at the crossroads of millennium development goals 4 and 5. Asia Pac J Public Health 450–463

Hirschman AO (1958) Strategy of economic development. Yale University Press, Yale

Hulme D, Wilkinson R (2012) The millennium development goals and beyond: global development after 2015. Routledge

Ibisch PL, Hoffmann MT, Kreft S, Pe'er G, Kati V, Biber-Freudenberger L et al (2016) A global map of roadless areas and their conservation status. Science 354(6318):1423–1427. Available on: https://doi.org/10.1126/science.aaf7166

IEA (2021) India energy outlook 2021: world energy outlook special report. International Energy Agency, Paris. Available on: https://iea.blob.core.windows.net/assets/1de6d91e-e23f-4e02-b1fb-51fdd6283b22/India_Energy_Outlook_2021.pdf

ILO (2018) Women and men in informal economy: a statistical picture, 3rd edn. International Labour Organization, Geneva. Available on: https://www.ilo.org/wcmsp5/groups/public/---dgreports/---dcomm/documents/publication/wcms_626831.pdf

ILO (2021) Young persons not in employment and education (NEET) in India: 2000–2019. International Labour Organization, New Delhi. Available on: https://www.ilo.org/wcmsp5/groups/public/---asia/---ro-bangkok/---sro-new_delhi/documents/publication/wcms_789243.pdf

ITU (2021) Individuals using internet (from any location by gender and urban/rural location %). International Telecommunication Union, Geneva. Available on: https://www.itu.int/en/ITU-D/Statistics/Pages/stat/default.aspx

Jacob N (2008) The impact of NREGA on rural-urban migration: field survey of Villupuram District, Tamil Nadu. CCS working paper no 202. Centre for Civil Society

Kadiyala S, Barnett T (2004) Econ Polit Wkly 39(19):1888–1892

Kapoor A, Sahoo RL (2004) Agricultutralists and the moneylender: a study of rural borrowing in Bihar and Orissa 1912–1936. In: Proceedings of the Indian history congress. Indian History Congress, pp 491–500. Available on: https://www.jstor.org/stable/44144764

Kaushal KK, Melkani VK (2005) India: achieving the millennium development goals through non-timber forest products. Int Forest Rev 128–134

Khera R, Nayak N (2009) Women workers and perceptions of the national rural employment guarantee act in India. In: Gender pathway out of poverty: rural employment. FAO-IFAD-ILO, Rome, pp 1–19

Kline P, Moretti E (2013) People, places, and public policy: some simple welfare economics of local economic development programs. Working paper 19659. NBER. Available on: https://www.nber.org/system/files/working_papers/w19659/w19659.pdf

Krishnan A (2018) Behind India's leap in ease of doing business. The Hindu, 11 Nov. Available on: https://www.thehindu.com/business/Economy/behind-indias-leap-in-ease-of-doing-business/article25469900.ece

Kroll C, Warchold A, Pradhan P (2019) Sustainable development goals (SDGs): are we successful in turning trade-offs into synergies? Palgrave Commun 5. Available on: https://doi.org/10.1057/s41599-019-0335-5

Kumar A, Valecha N, Jain T, Dash AP (2007) Burden of malaria in India: retrospective and prospective view. Am J Trop Med Hygiene 69–78

Lahiry S (2014) India's millennium development goals. World Aff J Int Issues 18(3):140–149

Loiwal M (2020) India Today, 31 Mar

Manning R (2009) Using indicators to encourage development: lesson from millennium development goals. DIIS. Danish Institute for International Studies, Copenhagen

Mathy S, Blanchard O (2016) Proposal for a poverty-adaptation-mitigation window within the green climate fund. Clim Policy 752–767. Available on: https://doi.org/10.1080/14693062.2015.1050348

Mattos FB, Dasgupta S (2017) MGNREGA, paid work and women's empowerment. International Labour Organization, Geneva. Available on: https://www.ilo.org/wcmsp5/groups/public/---ed_emp/documents/publication/wcms_613735.pdf

McMichael AJ, Friel S, Nyong A, Corvalan C (2008) Global environmental change and health: impacts, inequalities, and the health sector. BMJ 191–194

Mehrotra S, Sinha S (2017) Explaining falling female employment during a high growth period. Econ Polit Wkly 54–62

Meier GM (1976) Leading issues in economic growth. Oxford University Press, New York

Ministry of Education (2021) Initiatives by the school education sector in 2020–21 for continuing teaching and learning. Minstry of Education Government of India, New Delhi. Available on: https://dsel.education.gov.in/sites/default/files/update/covid_initiatives.pdf

Ministry of Rural Development (2011) Ministry of Rural Development, Government of India, New Delhi

mint (2016) Bihar extends 35% reservation for women in all govt jobs. mint, Jan 2016. Available on: https://www.livemint.com/Politics/jNlMwoB5KuaHUZX767EbeI/Bihar-extends-35-reservation-for-women-in-all-govt-jobs.html

Mohanty SK (2012) Multiple deprivations and maternal care in India. Int Perspect Sex Reprod Health 38(1):6–14

MOSPI (2017a) Millennium development goals—final country report. Central Statistics Office, Government of India, New Delhi. Available on: http://mospi.nic.in/sites/default/files/publication_reports/MDG_Final_Country_report_of_India_27nov17.pdf

MOSPI (2017b) Millennium development goals—final country report of India. Ministry of Statistics and Programme Implementation, Government of India, New Delhi

MOSPI (2021) Sustainable development goals national indicators framework: progress report 2021. Ministry of Statistics and Programme Implementation, Government of India, New Delhi. Available on: http://mospi.nic.in/sites/default/files/publication_reports/SDG-NIF-Progress2021_March%2031.pdf

Moyo S, Wright JA, Ndamba J, Gundry S (2004) Realising the maximum health benefits from water quality improvements in the home: a case from Zaka district, Zimbabwe. Phys Chem Earth 1295–1299

Nelson RR (1956) A theory of the low-level equilibrium trap in underdeveloped economies. Am Econ Rev 894–908. Available on: http://www.jstor.org/stable/1811910

Nerini FF, Julia Tomei LS, Bisaga I, Parikh P, Black M, Borrion A et al (2017) Mapping synergies and trade-offs between energy and the sustainable development goals. Nat Energy 10–15. Available on: https://www.nature.com/articles/s41560-017-0036-5

Neve J-ED, Sachs JD (2020) The SDGs and human well-being: a global analysis of synergies, trade-offs, and regional differences. Sci Rep 1–12. Available on: https://www.nature.com/articles/s41598-020-71916-9

Nilsson M, Griggs D, Visbeck M (2016) Policy: map the interactions between sustainable development goals. Nature 320–322. Avaoilable on: https://www.nature.com/articles/s41599-019-0335-5#ref-CR15

NIPUN BHARAT (2021) National initiative for proficiency in reading with understanding and numeracy: guideline for implementation. Department of School Education & Literacy, Ministry of Education, Government of India, New Delhi. Available on: https://www.education.gov.in/sites/upload_files/mhrd/files/nipun_bharat_eng1.pdf

NSSO (2014) Participation of women in specified: activities along with domestic duties. NSS 68th round. Ministry of Statistics and Programme Implementation, Government of India, New Delhi

NSSO (2015) NSS report no. 566: status of education and vocational training in India. Ministry of Statistics & Programme Implementation, New Delhi. Available on: https://mospi.gov.in/documents/213904/301563//nss_report_no_566_21sep151602577933947.pdf/a5443bf8-4a37-ae08-f4f5-85f125bd1a2d

NSSO (2020) Annual report: periodic labor force survey (PLFS). Ministry of Statistics and Programme Implementation, Government of India, New Delhi

Oommen OV, Laladhas KP (2016) Sustainable management and use of terrestrial ecosystem. In: RIS, India and sustainable development goals: the way forward. Research and Information System for Developing Countries (RIS), New Delhi, pp 181–194

Osborn D, Cutter A, Ullah F (2015) Universal sustainable development goals: understanding the transformational challenges for developed countries. United Nations, New York. Available on: https://sustainabledevelopment.un.org/content/documents/1684SF_-_SDG_Universality_Report_-_May_2015.pdf

Pangariya A, Mukim M (2014) A comprehensive analysis of poverty in India. Asian Dev Rev 1–52. Available on: https://documents1.worldbank.org/curated/en/847861468257945118/pdf/946460v10JRN0B0O0900170adev0a000021.pdf

Piot P, Greener R, Russell S (2007) Squaring the circle: AIDS, poverty, and human development. PLoS Med 4(10):e314, 1571–1575

Planning Commission (1985) Seventh five year plan: 1985–1990. Planning Commission of India, New Delhi

Poalnki P (2012) How UP and Bihar are skewing India's poverty estimates. First Post, 20 Mar. Available on: https://www.firstpost.com/business/economy/how-up-and-bihar-are-skewing-indias-poverty-estimates-250226.html

Poku N, Whitman J (2011) The millennium development goals: challenges, prospects and opportunities. Third World Q 3–8

Pradhan P, Luís Costa DR, Kropp JP (2017) A systematic study of sustainable development goal (SDG) interactions. Earth's Future 1169–1179. Available on: https://doi.org/10.1002/2017EF000632

Purvis K (2016) 12 ways environment and development sectors can collaborate to meet the SDGs. The Guardian, 31 Oct

Radkar A, Parasuraman S (2007) Maternal deaths in India: an exploration. Econ Polit Wkly 3259–3263

Ravallion M, Chen S (2003) Measuring pro-poor growth. Econ Lett 93–99

Ravallion M, Datt G, Walle DV (1991) Quantifying absolute poverty in the developing world. Rev Income Wealth 345–361

Ravindranath NH, Murthy IK (2021) Mitigation co-benefits of carbon sequestration from MGNREGS in India. PLoS ONE

RBI (2021) Table 10: per capita net state domestic product—state-wise (at constant prices). Reserve Bank of India, New Delhi. Available on: https://www.rbi.org.in/scripts/PublicationsView.aspx?id=20415

Registrar General of India (2011) Censul 2011. Registrar General of India, Government of India, New Delhi

Robert, Fioramonti, Kubiszewski I (2016) The UN sustainable development goals and the dynamics of well-being. Front Ecol Environ 59. Available on: https://doi.org/10.1002/fee.1231

Roberts K (1980) Price-independent welfare prescriptions. J Public Econ 13:277–297. Available on: https://www.sciencedirect.com/science/article/abs/pii/0047272786900071

Samuelson PA, Swamy S (1974) Invariant economic index numbers and canonical duality: survey and synthesis. Am Econ Rev 566–593. Available on: http://www.jstor.org/stable/1813311

Scherer L, Behrens P, Koning A, Heijungs R, Sprecher B, Tukker A (2018) Trade-offs between social and environmental sustainable development goals. Environ Sci Policy 65–72. Available on: https://doi.org/10.1016/j.envsci.2018.10.002

SECC (2011) SECC 2011 at a glance. Ministry of Rural Development, Government of India, New Delhi. Available on: https://secc.gov.in/getSeccDataSummaryNationalReport.htm

Sen A (1976) Poverty: an ordinal approach to measurement. Econometrica 219–231

Sen A (1979) Equality of what? The Tanner lecture on human development, 22 May, pp 195–220

Sen A (1999a) Development as freedom. Knopf, New York

Sen A (1999b) Development as freedom: an India perspective. Indian J Ind Relat 157–169

Sen A (2001) Many faces of gender inequality. Frontline. Available on: https://www.sas.upenn.edu/~dludden/MANY%20FACES%20OF%20GENDER%20INEQUALITY.htm

Sen A (2005) The three R's of reform. Econ Polit Wkly 1971–1974. Available on: http://www.jstor.org/stable/4416610
Sen A, Dreze J (2011) Putting growth in its place: it has to be but a means to development, not an end in itself. Outlook, 14 Nov. Available on: https://www.outlookindia.com/magazine/story/putting-growth-in-its-place/278843
Shah M (2007) Employment guarantee, civil society and Indian democracy. Econ Polit Wkly 43–51. Available on: http://www.jstor.org/stable/40276832
Sharma R (2018) Health and economic growth: evidence from dynamic panel data of 143 years. PLoS ONE. https://doi.org/10.1371/journal.pone.0204940
Sharma AN, Saluja MR, Sharma A (2016) Macro economic impact of social protection progrrames in India. Econ Polit Wkly 121–126
Singh M (2012) PM's opening remarks at the meeting of PM's national council on skill development. Prime Minister Office, New Delhi. Available on: https://archivepmo.nic.in/drmanmohansingh/content_print.php?nodeid=1126&nodetype=2
Singh UK (2021) India has a serious hunger problem and it needs urgent policy intervention. The WIRE, 30 Oct. Available on: https://thewire.in/food/india-has-a-serious-hunger-problem-and-it-needs-urgent-policy-intervention
Singh NK, Stern N (2013) The new Bihar. HarperCollins
Smith P, Olesen JE (2010) Synergies between the mitigation of, and adaptation to, climate change in agriculture. Agric Sci 543–552. Available on: https://doi.org/10.1017/S0021859610000341
Tewary A (2020) Tejashwi Yadav's apology for 'mistakes' of RJD regime fails to impress Bihar's ruling NDA. The Hindu, 3 July. Available on: https://www.thehindu.com/news/national/other-states/tejashwis-apology-for-mistakes-of-rjd-regime-fails-to-impress-bihars-ruling-nda/article31982878.ece
Thapar R (1966) A history of India. Penguin Books, London
Thomas D, Frankenberg E (2002) Health, nutrition and prosperity: a microeconomic perspective. Bull World Health Organ 106–113
UN (1972a) Report of the United Nations conference on the human environment. United Nations, New York. Available on: https://undocs.org/en/A/CONF.48/14/Rev.1
UN (1972b) United Nations conference on the human environment, Stockholm, 5–16 June 1972. United Nations, New York
UN (1987) Report of the world commission on environment and development: our common future. United Nations, New York. Available on: https://sustainabledevelopment.un.org/content/documents/5987our-common-future.pdf
UN (1992a) Report of the United Nations conference on environmemt and development, vol I. United Nations, New York. Available on: https://undocs.org/en/A/CONF.151/26/Rev.1
UN (1992b) United Nations conference on environment & development: agenda 21. United Nations, New York
UN (1992c) United Nations conference on environment and development. United Nations, New York
UN (2002) Report of the world summit on sustainable development. United Nations, New York. Available on: https://undocs.org/pdf?symbol=en/A/Conf.199/20
UN (2012a) Resolution adopted by the General Assembly on 27 July 2012. United Nations, New York. Available on: https://www.un.org/ga/search/view_doc.asp?symbol=A/RES/66/288&Lang=E
UN (2012b) Resolution adopted by the General Assembly on 27 July 2012. Agenda No. A/RES/66/288. United Nations, New York
UN (2013) Open working group of the General Assembly on sustainable development goals. Agenda No. A/67/L.48/Rev.1. United Nations, New York
UN (2015a) Resolution adopted by the General Assembly on 25 September 2015. A/RES/70/1. United Nations, New York

UN (2015b) Resolution adopted by the General Assembly on 25 September 2015. Agenda No. A/70/L.1. United Nations, New York. Available on: https://www.un.org/ga/search/view_doc.asp?symbol=A/RES/70/1&Lang=E

UN (2015c) When we stand together, there is no limit to what we can achieve, Ban tells UN Assembly. United Nations, New York. https://www.un.org/sustainabledevelopment/blog/tag/secretary-general-ban-ki-moon/

UN (2016a) MDG monitor: MDG 6: combat HIV/AIDS, malaria and other major diseases. United Nations, New York

UN (2016b) Report of the inter-agency and expert group on sustainable development goal indicators. Economic and Social Council, United Nations, New York. Available on: https://undocs.org/E/CN.3/2016/2/Rev.1

UN (2017) The sustainable development goals report: 2017. United Nations, New York. Available on: https://unstats.un.org/sdgs/files/report/2017/thesustainabledevelopmentgoalsreport2017.pdf

UN (2020a) Policy brief: the impact of COVID 19 on children. United Nations, New York. Available on: https://unsdg.un.org/sites/default/files/2020-04/160420_Covid_Children_Policy_Brief.pdf

UN (2020b) The sustainable development goals report: 2019. United Nations, New York. Available on: https://unstats.un.org/sdgs/report/2019/The-Sustainable-Development-Goals-Report-2019.pdf

UN (2020c) United Nations in India. United Nations, New York. Available on: https://in.one.un.org/page/sustainable-development-goals/gender-equality-sdg-5/

UN (2021) The sustainable development goals report: 2021. United Nations, New York. Available on: https://unstats.un.org/sdgs/report/2021/

UNAIDS (2021) Global HIV & AIDS statistics—fact sheet. Retrieved from: https://www.unaids.org/en/resources/fact-sheet

UNDP (1997) Human development report 1997. United Nations Development Programme, New York

UNDP (2015) Building the post-2015 development agenda: 2014 annual report. United Nations Development Programme, New York. Available on: https://www.undp.org/sites/g/files/zskgke326/files/publications/UNDP_Building%20the%20Post-2015%20Development%20Agenda_Final_2015.pdf

UNDP (2018) Human development report 2018. United Nations Development Programme and Oxford University Press, New York

UNDP (2019) Global multidimensional poverty index 2019: illuminating inequalities. United Nations Development Programme and Oxford Poverty and Human Development Initiative, New York. Available on: http://hdr.undp.org/sites/default/files/mpi_2019_publication.pdf

UNICEF (2016) A snapshot of Bihar elementary education: district-wise statistical report. UNICEF BIHAR, Patna. Available on: https://www.bepcssa.in/en/A%20Snapshot%20of%20Bihar%20Elementary%20Education1.pdf

UNICEF (2017) A world free from child poverty: a guide to the tasks to achieve the vision. United Nations Children's Fund, New York

UNICEF (2020a) Maternal mortality in 2000: estimate developed by UNICEF and UNFPA. United Nations Children's Fund, New York

UNICEF (2020b) Universal child benefits: policy issues and options. UNICEF, New York. Available on: https://www.unicef.org/media/72916/file/UCB-ODI-UNICEF-Report-2020.pdf

UNICEF India (2016) Children in India: India's growth over the last two decades has contributed phenomenally to global human development. United Nations Children's Fund, New Delhi

Wheebox (2021) India skills report 2021. Wheebox, Gurugram. Available on: https://indiaeducationforum.org/pdf/ISR-2021.pdf

WHO (1990) Maternal mortality in 2000: estimate developed by UNICEF and UNFPA. World Health Organization, Geneva

WHO (1991) Maternal mortality: a global factbook. World Health Organization, Geneva

WHO (2019) Maternal mortality. World Health Organization, Geneva

WHO (2020) Global report on WASH in health care facilities. World Health Organization, Geneva

World Bank (1993) World Bank development report 1993: investing in health. The World Bank Group & Oxford University Press, Washington, DC

World Bank (2008) World development report 2008: agriculture for development. The World Bank, Washington, DC. Available on: https://doi.org/10.1596/978-0-8213-6807-7

World Bank (2009) Doing business in India 2009. The World Bank, Washington, DC. Available on: https://openknowledge.worldbank.org/handle/10986/28088#:~:text=Doing%20business%20in%20India%202009%20is%20the%20first%20country%20specific,cities%2C%20and%20documents%20their%20progress

World Bank (2017) New World Bank project to give impetus to India's skill's agenda—to provide over 8 mn youth with market-relevant training. The World Bank Group, Washington, DC

World Bank (2020) Global economic prospects, June 2020. The World Bank, Washington, DC. Available on: https://openknowledge.worldbank.org/handle/10986/33748

World Bank (2021) World Bank indicators. The World Bank, Washington, DC. Available on: https://data.worldbank.org/indicator/SP.DYN.IMRT.IN

Worldometers (2020) Coronavirus cases. Worldometers. Available on: https://www.worldometers.info/coronavirus/coronavirus-cases/?fbclid=IwAR1rXCB1NYf48HuNqhHZDPNSUySwz8PpWwqaGRsE-3PijhQfc03axd2FPcI

Zhao Z, Cai M, Wang F, Winkler JA, Connor T, Chung MG, Zhang J, Yang H, Xu Z, Tang Y, Ouyang Z, Zhang H, Liu J (2021) Synergies and tradeoffs among sustainable development goals across boundaries in a metacoupled world. Sci Total Environ 1–18. Available on: https://doi.org/10.1016/j.scitotenv.2020.141749

Chapter 7
Conclusion and Way Forward

Abstract This concluding chapter provides an outline of the progress achieved by India and Bihar in fulfilling human development aligned with development programmes framed under the aegis of UNDP. It highlights the necessity of synergizing efforts and collaboration toward achieving the ends. Partnership with international agencies such as UNDP is critical to learn from the best practices across the globe.

Keywords Bihar · India · Public policies · HDI · MPI

7.1 Introduction

The MDGs and targets represent a supremely ambitious and transformational vision. It envisages a world free from poverty, hunger, disease, and want, where all life can flourish. MDGs envision a world where every individual attained universal primary education. A world with equitable and universal gender rights that are protected at all levels, where children receive proper health care and social protection, and where physical, mental, and social well-being are assured. It reaffirms our commitments regarding the human right to safe drinking water and sanitation and where there is improved hygiene and safe, affordable, nutritious, and sufficient food. MDGs visualize a world where global partnership leverages economic growth.

At the end of fifteen years time frame, the MDGs have made an important and significant progress in a number of areas. The MDGs remain off-track, in particular those related to poverty, maternal, newborn, and child health reproductive health and gender equality. UNDP implements SDGs to recommit to the full realization of all the unfinished goals of MDGs, including the off-track, particularly by providing focused and scaled-up assistance in line with relevant support programs. The new agenda seeks to attain what these did not achieve, in reaching the most vulnerable.

Inferred from Chap. 2, the MDGs' key targets were halving poverty, fortifying universal primary enrolment, ensuring attainment of gender parity, reducing the maternal mortality rates by three-fourths, curbing the child mortality by two-thirds, and decreasing the incidence of HIV/AIDS, among others. In the year 2000, India as a member of the UN became a signatory of this important journey to restore the dignity

U. K. Singh, *Human Development in Bihar, India*, Advances in Geographical and Environmental Sciences, https://doi.org/10.1007/978-981-19-3624-1_7

of humanity and the environment. As part of this international movement, India was determined to play its role in reducing extreme poverty and other forms of present disadvantage which have been persecuting all the societies for centuries. Sometimes despite public pessimism, in the past decades, India has witnessed different improvements in the lifestyle of its citizens.

Statistical evidence (as discussed in Chaps. 2 and 3), reflects that India has made stimulating improvements in the reduction of extreme poverty. Forward-looking schemes like MGNREGA, Standup India, Skill India, and Startup India have played a vital role in uplifting the people from poverty. However, poverty levels among marginalized sections such as women and children still remain a challenge.

7.2 Achievement During MDGs

In India, "the number of people living with extreme poverty has changed by more than half, slipping from 1.9 billion in 1990 to 836 million in 2015, resulting in the attainment of goal before the target year". The Poverty Gap Ratio (PGR) also declined at the national level. However, the rural PGR declined from 9.64 in 2004–05 to 5.05 in 2011–12 in the urban areas it declined from 6.08 to 2.70 during the same period. At the state level, Bihar has done well, as depicted in Fig. 7.1, the PHCR for Bihar is 60.5 in 1993–94, comes to 33.74 in 2011–12 likely to achieve the target of MDG 1 while at national level India has achieved the target in 2011–12 (MOSPI 2015).

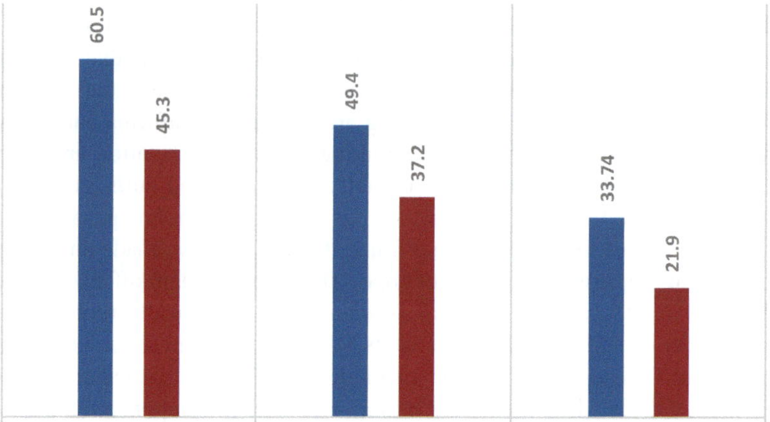

	1993-94	2004-05	2011-12
Bihar	60.5	49.4	33.74
India	45.3	37.2	21.9

Fig. 7.1 Trend of PHCR of India and Bihar. *Source* MOSPI (2015)

India has also performed well on the front of MDG 1, during the past decades and half. India witnessed poverty reduction by lifting its deprived population from acute poverty. Even so, it is home to many people living in disgusting poverty. The country is facing the challenges in form of price rises, low buying capacity, and insufficient social structure affecting the living standard and job prospects (MOSPI 2015). The Government of India runs various programs like MANREGA, NRLM, and many more as discussed in Chaps. 2 and 3 are generating employment vis-à-vis raising the purchasing power to uplift their citizens from extreme poverty.

Inferred from Chap. 2, the country has achieved the target of universal primary enrolment. The right to education is the fundamental right in both countries. In India, high enrolment rates reflect the success of programs like "Sarva Siksha Abhiyan and Mid-Day Meal" responsible to provide a fillip to the "Net Enrolment Ratio" in India, especially in Bihar. As the net enrolment ratio in 2006 is 26.5% which comes to 43.37% in 2009 but it climbed to 93.77 in 2014. This is the result of achieving universal primary education. The quality of education also improved with the increase in the number of qualified teachers, the improvement in LER and by strengthening of school infrastructure.

Education leverage women empowerment, women can realize their personal ambition and career prospects through the attainment of education. Gender equality ensures women's empowerment with a focus on identifying and redressing power imbalances and by making them more autonomous to handle their own lives. Women empowerment comes with so many benefits such as they became a sole contributor to their families by getting better job opportunities and thus benefiting the entire family. It is important to establish societies where rights and opportunities are not affected by gender.

In India, the rate of literate women (15–24) has shown an upward trend (rural and urban). During the period of 1991–2011, increase in literacy trends is registered among female youths from 49.3% in 1991 to 81.8% in 2011. Bihar with various schemes like "Balika Cycle Yojana", "Balika Poshak Yojana", and many more (as discussed in Chap. 3) is improving the gender parity index, it was 0.08 till 2011.

The pace of increasing the number of women in parliament and legislative assembly is very slow in India. Figure 7.2, reflects the affirmative trend of women's representation in Indian parliament (Lok Sabha and Rajya Sabha), and it was 9.7% in 1991 to 12.24% in 2015. The Bihar government has given 50% reservation

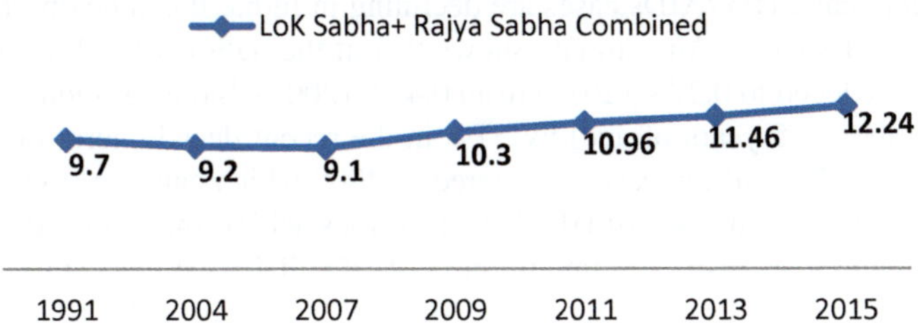

Fig. 7.2 Trend of women participation in Indian Parliament. *Source* Lok Sabha and Rajya Sabha

Table 7.1 Proportion of 1-year-old children immunized against measles percentage in India

Country	2003	2008	2012	2104
India	58	59	74	75
Bihar	68	76.9	73	75

Source WHO (2015)

to the women candidate at the local bodies' election are enjoying the benefit by implementing a law to uplift the women in their respective constituencies.

With the improved coverage of the health system, immunization, breastfeeding, and the use of integrated management of childhood illness method is responsible for reducing mortality. Infant mortality was reduced by 61 deaths per 1000 in India.

Table 7.1 illustrates the improvement in measles vaccination coverage, that is, increasing from a baseline of 74.2% in 2003 to 91.2% in 2014 in India from 58 to 72% in the same period. On the provincial base, Gauteng registered a stride rate of 68.8% to 106.8 while in the same period Bihar registered an improvement of 68–75%.

Poverty and preference for male children and early marriage of girls are the biggest challenges for infant mortality in India. The government recognized the challenge and through its program "Beti Bachao, Beti Padhao" (preserve Girl Child, Educate Girl Child), this mission aims to rejoice the girl child delivery and ensure better education as par with boys.

To balance the sex ratio, the government has put a bounty of Rs. 1 crore, for the "innovative" village which attains a balanced sex ratio. "Sukanya Samriddhi" account (girl child prosperity scheme), is the program under which girl children below 10 years will have bank accounts with 9% interest and income tax benefits and it will remain active till 21 years of age.

The MDG 5 target for India stipulates reducing MMR by three-fourths, between the periods of 1990–2015. 'In the base year 1990, the estimated MMR was 437/100,000' (MOSPI 2017c). To overcome the situation India is strengthening the health structure of the country to order to accomplish the MDGs.

In India, the institutional deliveries registered 82.86% in 2013 from 70.625% in 2009. This evidence suggests that the awareness among the people for safe deliveries as well as strengthening of the health system in both the countries is in the right direction.

In recent years, HIV/AIDs cases are declining in India. It can be observed in all the sections of society. An estimate shows that at the national level occurrence of HIV/AIDS reduced to 0.27% (2011) from 0.41% (2001). India has reduced the HIV infection (above 15 years of age) by 57% in the recent decade, term of a number of cases was 0.274 million cases registered in 2000 which came to 0.16 million in 2011. The reduction in cases of HIV is a marvelous achievement for India.

TB incidences are also declining in India. WHO (2015) estimates that the occurrence of TB cases has been reduced to 211/100,000 (2013) from 465/100,000 (1990). Cases of TB are declining from 216/100,000 (2011) to 171/100,000 (2013). Deaths due to TB are reduced to 19/100,000 (2013) from 38/100,000. Trends suggest that TB goals are accomplished by India.

India has also done well in ensuring household access to improved drinking water. At the national level data shows a positive trend as 87.8% of household got a better supply (rural: 86.9 urban: 90.1%).

India is far behind in the availability of latrines. At the national level 43.4% of people do not have access to a latrine. The government is concerned about the improvement of water and sanitation programmes through programs like "National Rural Drinking Water Programme", "Swachh Bharat Abhiyan (Gramin)", "Nirmal Gram Puraskar" push the backlog.

The idea of development is always underscored by a certain degree of measure of underdevelopment which is done away with. In the juxtaposition to the case of Bihar and India, we must follow similar routes to follow development. The onus is on the Indian and Bihar governments to revamp the health sector and bring down the female mortality rates and also focus on bettering child health and child care. The non-organized and non-formal sectors must be taken more seriously so that the participants must feel empowered at all levels. The skill-based labor is the requirement of the day for both the countries and they could serve as a model for rest of underprivileged areas to follow. The skill-based education is vital for the socioeconomic growth of the country. It enables better job opportunities, health care, and reduces inequalities. The people can develop gainful skills to be employable as per the market scenario. Due to higher employment opportunities, people can afford better health facilities. A higher level of disposable income in the hand of people at large is helpful to reduce inequality. The skill should be developed in children from the young stage. It is also the need of time to strengthen the free and compulsory education by introducing compulsory skill training. It increases the job prospects vis-à-vis accelerating the economic growth of the country. Such a system has been adopted by developing countries for a long time. It is high time for India to follow the same path of development.

The measures of development must not only focus on reserving seats for women to empower them but also the focus of the policies should be to bring women to the forefront in power relations. The women should not be assisting the men in development whereas they should be equal partners and more. The key is to bring them on a decision-making platform.

All social policies such as MGNREGA, Antodaya, Sarva Shiksha Abhiyan, Midday Meal, Maatritva Suraksha, Yojana, Jeevika, etc. must reshape themselves to cater to the needs of the subjugated and the left-outs in the society. We have got economic and political transformations; all we need today is to achieve Social Transformations.

7.3 Conclusion

The Millennium Summit (2000) was a path-breaking measure to counter the problems like poverty, hunger, and gender inequality; and the world leaders agreed to the development goals. Such goals which were agreed upon by the world leaders were measurable as well as specific in nature, with an objective to bring about the change. India was one of the signatories to the Millennium Declaration and had consistently worked toward the goals, in fact, the country converged its development goals to align with the Millennium Development Goals (MDGs). One of the significant characteristics of such an initiative was its influence on the planning and policy formulation globally. According to the UNDP, the MDGs brought about the critical development challenges to the front and also facilitated the countries with a strong target-oriented agenda. As 2015 approached, the success as well as the failure of this initiative was brought to light, while many people have been elevated from poverty and deprivation, many others continue to stay challenged with deprivation. Many objectives have been achieved by India, like gender parity has been achieved in the primary school enrollment, providing clean drinking water, etc. yet the other objectives like sanitation, primary school enrollment, and its completion remained inadequate.

The failure to produce the desired results witnessed a collapse as the successes were not equally distributed across the globe. For instance, the same objective of Poverty Reduction by 50% was defeated by the regions like parts of Asia, Africa, etc. The statistics have presented that Southeastern Asia exceeded the goal of poverty reduction by 16%, Southern Asia exceeded the goal by 12.5%, Northern Africa dragged by about 1.2%, sub-Saharan Africa was the last as it did not even meet the goal of extreme poverty reduction, as it stood 12.5% away from the goal.

The objective of the extreme poverty reduction to 50% reduction for those living at $1.25 a day arguably had the best statistics for each country, from there it witnessed a collapse in performance. This trend was seen in many other Millennium Development Goals, sub-Saharan Africa was far behind in such an endeavor, while all the countries failed to achieve the goal set for the maternal mortality rate reduction. For the gender inequality objective, the initiatives undertaken involved the efforts like more representation of women in parliament, and more girls going to school. However, women continue to face the impediments and challenges in access to work, economic assets, and in private and public decision-making; according to the UN.

Further as quoted by U.N., "children from the poorest 20 percent of households are more than twice as likely to be stunted as those from the wealthiest 20 percent and are also four times as likely to be out of school. In countries affected by conflict, the proportion of out-of-school children increased from 30 percent in 1999 to 36 percent in 2012".

In addition, the figures for the global carbon emissions and the water scarcity are alarming as well. Carbon emissions have increased by 50% while water scarcity affects more than 40% of the world when compared to 1990 statistics. In such a scenario, all the hopes are only from the Sustainable Development Goals. "The emerging post 2015 development agenda, including the set of Sustainable Development Goals, strives to build on our successes and put all countries, together, firmly on track towards a more prosperous, equitable and sustainable world", said by Ban Ki-moon, U.N. General Secretary.

The performance of India in terms of achieving the Millennium Development Goals (MDGs) has been inconsistent, with some targets achieved ahead of the 2015 deadline. The MDG 1, which aimed at eradicating extreme Hunger and Poverty in India has been achieved moderately by India. The Poverty Head Count Ratio (PHCR) stood at 47.8% in 1990, and although the target for India for the year 2015 was 23.9%, the PHCR for the country in 2011–12 was 21.9%. Although such progress has been uneven, a major factor in achieving such an objective was the faster reduction in poverty since the mid-2000s. The eradication of hunger continues to be a challenge for the country. With over a third of the world's underweight children and near a third of the world's food-insecure people. The MDGs were formulated in the 1990s when 53.5% of all Indian children were malnourished. Malnourishment declined to 40% in 2015, which is still away from the target of reducing malnourishment to 26%.

The second goal to achieve Universal Primary Education was effective, with enrollment and the completion rates of primary school education for girls as well as elementary completion rates reaching close to those of the boys. India has only 65 women representatives out of 542 members in Lok Sabha, and 31 female representatives in the 242 members in Rajya Sabha. The proportion of the women in National Parliament is only 12.24% against the target of 50%.

The third goal to reduce Child Mortality was achieved moderately, with a decline in the Under-Five Mortality Rate (U5MR) from 125 per 1000 live births in 1990 to 49 per 1000 live births in 2013. Although the target under MDGs was 42 per 1000, the moderate achievement highlights a few areas like—neonatal and childhood illnesses, child survival needs to be worked upon (particularly for vulnerable communities).

The fifth goal was to Improve Maternal Health and this particular goal witnessed a sluggish and flat growth. From 1990 to 1991, the Maternal Mortality Rate (MMR) was 556 per 100,000 live births, and the year 2009 recorded an MMR of 167 per 100,000 live births, a slight improvement.

The sixth goal was to Combat HIV/AIDS, malaria, and other diseases, a progress could be observed with HIV, tuberculosis, malaria cases declining. The HIV/AIDS cases prevalence was particularly due to sexual activity. Such adult prevalence declined from 0.45% in 2002 to 0.36% in 2009. This decline could be achieved due to greater awareness and increased use of condoms. Further for the cases of tuberculosis (TB), praiseworthy initiatives were undertaken such that tuberculosis prevalence per lakh population reduced to 211 in 2013 from 465 in the year 1990. TB incidence per 100,000 population declined from 216 in 1990 to 171 in 2013.

The seventh goal to Ensure Environmental Sustainability was achieved in terms of an increase in Forest Cover to 21.23%—an increase of 5871 km^2. The overall proportion of households having access to improved water sources increased from 68.2% in 1992–93 to 90.6% in 2011–12. Another indicator is the sanitation coverage, the country has recorded very slow progress in that area.

The eighth goal to Develop a Global Partnership for Development witnessed a huge leap in the telecom sector and IT sector advances. UNDP (2018) report explained that India has 946.4 million, including the 918.72 million wireless ones; the Indian Telecom Network is the second-largest network in the world after China. Out of this, 383.97 million telephone connections are in rural areas and 562.43 million are in urban areas. The respective numbers for the IT sector have also witnessed an increase, the Internet subscribers have increased from 198.39 to 259.14 million from 2013 to 2014, with an annual growth of 60%.

The success and failures in terms of achievement are uneven for the different states of India. The state of Bihar is a key area to study as it is the least urbanized state in India. The state has an urban population of about 10%. The population employed in the agriculture sector is 73% of its workforce. The per capita agricultural income is half of that of India. However, the state paints a gloomy picture of its under-performance in terms of generation of employment in the nonagricultural sectors when compared to the other states. The rural industries like cotton, jute, etc. have witnessed a downfall. The few changes observed in the present situation of Bihar can be accrued to the attention it has received in the past 4–5 years, and this comes as a major breakthrough for the state that has been in the state of perpetual poverty for decades.

Thus, the performance of Bihar is critical to the evaluation of the performance of India toward the achievement of the Millennium Development Goals (MDGs). The first goal which deals with the eradication of poverty has witnessed a plausible improvement. The state of Bihar has succeeded to bring down the Poverty Headcount Ratio (PHCR) from 49.4% in the year 2004–05 to 33.74% in the year 2011–12. However, the state shall continue with its efforts towards the fight to bring out 35.8 million people who continue to live in poverty and deprivation.

The second goal which was an endeavor toward education can be deemed as a major achievement with the literacy rate rising from 47% in 2001 to 61.8% in 2011.

The third goal aspires to promote gender equality and women empowerment. Gender equality demands the empowerment of women, the focus of such an objective is to identify and redress the power imbalances and give women more autonomy to manage their own lives. When women are empowered, the whole family benefits, thus the society as a whole benefits thereby creating ripple effects on future generations. The area has seen a progress between the years 2001 and 2011 census. The female literacy for the year 2001 stood at 51.5% against the 33.1% in the year 2011. The decadal increase in female literacy (recorded as 18.4%) is much higher than the decadal growth of male (11.5%) (Census 2001, 2011). Further during the decade 2001–2011, the percent increase in the female youth literacy to male youth literacy ratio was 17%, and the growth was 13.75%. The indicator presented a promising future.

The fourth goal was to Reduce Child Mortality, Bihar is the most populous state which accommodates the second largest population in the country. The target was missed by the state of Bihar narrowly. The target under the fourth goal was to be achieved by reducing the Under Five Mortality Rate (U5MR) to 42 deaths per 1000 live births by 2015, while as per the historical trends the target was predicted to be missed narrowly with 48 deaths per 1000 live births.

The fifth goal was to Improve Maternal Health (MMR), the goal as per the Millennium Development Goals (MDGs) was to bring the levels to three-fourths between 1990 and 2015. In 1990, the estimated MMR was 437 per 100,000 live births. Such a target could be achieved when the MMR was reduced to 109 per 100,000 live births in the year 2015. The Maternal Mortality Rate (MMR) for 2014–16 for Bihar/Jharkhand is 165 per 100,000 live births (NITI Aayog 2019). The target has been largely missed by the state of Bihar as it moves toward the Sustainable Development Goals.

The sixth goal was to combat HIV/AIDS, Malaria, and other diseases; the goal can be analyzed with the help of its components. The first component is the HIV prevalence, while the nationwide HIV-prevalence levels declined, Bihar was among the few states of India which witnessed a reverse trend in terms of HIV-prevalence including states like Jharkhand, Punjab, Assam, Uttarakhand, and Odisha. Such a trend could be attributed to the availability of better test facilities. Further, the other indicator like the condom use rate has improved from 2.3 to 3.1% during NFHS-4 (2015–16). The second component is malaria disease, Bihar has taken several initiatives to contain the vector-borne disease—Malaria. The third component is Tuberculosis; the state of Bihar has witnessed as many as 58,814 new cases of TB in 2014. The cases which were recorded include 1010 cases of multi-drug resistant (MDR) TB and 25 cases are of extensive drug-resistant TB, known as XDR [Source: TOI March 23, 2015]. The TB incidence fell in India, the country has met the target 50% reduction in prevalence rate by 2015. Although for the state of Bihar collectively, the goal remained unachieved.

The seventh goal was environment and sanitation, with the view of integrating the principle of sustainable development into the country's policies and programs and reversing the loss of environmental resources. There were promising changes witnessed for the state of Bihar, the forest cover expansion rate for the state was 0.47% during 2007–2013; the figures at the national level were only 0.18% (ISFR 2013).

Integration of the world provided an opportunity to build a platform for global partnership, the idea resonated in the call of MDG 8, partnership for development. Cooperation between the countries opens the avenue to learn from the expertise of each other, by the mean of globalization, driven by information and communication technology (ICT). In the integrated world, ICT is recognized as an infrastructure base required for enablers of economic growth (ADB 2010). World Bank (2013) suggests that ICT has potential to accelerate the economy in developing countries by inviting investment and creating employment opportunities.

Benefitting the rural area with the advancement of ICT is a difficult task. Tele density in India is uneven depending on the geographical and socioeconomic status of people. Rao's (2007) finding reflects that Bihar with teledensity of 0.64 fell behind the national average of 1.81, which is improved during 2016 (0.54) and 2018 (0.63), and the state became second in terms of urban teledensity in India (TNN 2019), the improvement gives the strong indication of economic growth in Bihar.

Goals and targets of SDGs were under-performed in the report based on SDGs index performance. Alkire and Foster (2011), developed an idea to measure the various dimension of poverty, which is called the multidimensional poverty index (MPI). This approach broadened the concept of human development by adding new indicators to the existing three initiators of the Human Development Index (HDI). MPI covered almost all the aspects concerning human development. The ten indicators are an expansion of HDI indicators, health (nutrition and child mortality); education (number of years in school and school attendance); standard of living (house chores fuel, sanitation, clean drinking water, affordable electricity, cemented house roof, and asset accumulation). If any individual does not satisfy at least one-third of the indicators, he shall be considered poor.

The MPI report (2019), revealed the misery of multidimensional poverty in Bihar. An estimated 52.2% of people of Bihar live in multidimensional poverty. The district of North Bihar is more deprived, these districts accommodate most of the poor population of Bihar. Evidence shows that the proportion of people with MPI reaches 60% in these districts; the number accounted for in Araria and Madhepura was 7 people with MPI out 10 people. Altogether these districts are home to 28 million poor from an overall 40 million of India's poor population (Singh 2020).

The health sector of Bihar is in a sorry state. Bihar is losing their children's life due to a lack of health facilities. NITI Aayog report (2019) reinstates Bihar as second on the ranking ladder of the country, trailed by Uttar Pradesh. The report sheds light on the health aspect of other age groups with respective Infant Mortality Rate (38/1000), which is the highest among all the states, and above the national average of 34/1000 (male + female). The Global Nutrition report (2020) found malnutrition as the major cause of infant mortality; it is responsible for 68% of Under-5 mortality. NFHS-5 survey report for 18 Indian states revealed that 35% of under-5 children are stunted, 30% are wasted, the proportion is highest in caparison to other parts of world.

The phenomenon observed shows a decline in the rates of stunting and wasting as the countries move upward on the income ladder. However, India is an outlier and breaks such a causality with its situation even worse than the Sub-Saharan African countries. The state of Bihar falls in the category of states like Manipur and West Bengal with a similar per capita income ($) as sub-Saharan African countries like—Liberia, Tanzania, and Zimbabwe—but, the average stunting rates in Bihar (43%), Assam (35%), and West Bengal (34%) are 10–12% higher than that of Liberia (33%), Tanzania (32%) and Zimbabwe (3%), respectively. There are lopsided measures undertaken on the part of the country. However, a study undertaken by the WHO attributes stunting to the medical and socioeconomic factors. The medical factors include genetics (parent's height), access to nutrition, and mother's health (anemia and BMI) while the income, poverty, access to healthcare, mother's education and labor force participation fall into the category of economic factors; and the social factors include caste, race, women status, place of residence, etc. Of these factors which one is proximate and which one is a distant factor is not known. The WHO recommendations for the doctor are one for every thousand while the state of Bihar lags with one doctor to serve 43,788 people. This further adds to the required initiatives in the field of healthcare to reap the benefits of demographic dividends for the states.

The demographic advantage of Bihar makes it India's youngest state where the mean age is 20 years. The Census (2011) reported that 61.5% of the total population is employed, and most of them depend on agriculture for their livelihood. In numbers, out of the total employed population of 34,724,987; the number of cultivators was 5,413,181; agricultural laborer constitutes 9,537,418. In search of better job opportunities, large chunk of Bihar's population migrated to other states. Labor of Bihar migrated to other states of India and Gulf countries. The two states of India have experienced the largest outside migration, 2.6 million from Uttar Pradesh and 1.7 million from Bihar (Census 2011). A combined report of Aajeevika and Tata Institute of Social Science Research (2017) finding revealed that the migrants from Bihar are employed in blue-collar jobs, mostly in construction, mining, agricultural labor, and any low skilled jobs. Share of Bihar's rural migration to urban cities of India constitutes only 12%, other 79% are engaged in low-skilled activities. The labor force participation in Bihar is 41.5%, below the national average of 54.5% (PLFS 2020).

The Socio-Economic Census—2013 talked about the homeless, and Bihar held second the last position in sheltering the homeless people followed by the West Bengal. The report pegged 3,402,635 as the number of homeless people in Bihar out of a total of 4.3 crore homeless people in the country.

Bihar is a revenue deficit state, hence it is unable to establish big institutions but with political cooperation—initiatives like small skill centers to impart decent jobs can help in creating opportunities for the youth, in a globalized world. There are economists who believe that the state needs a developmental model like Bangladesh, considering the fact that the two are very similar in terms of demography; moreover, the latter is also one of the fastest-growing economies in Asia. The implementation of such a model implies a concentration on the value addition to succeed, on part of Bihar.

7.4 The Way Ahead

Bihar situated in the eastern part of India share boundaries with West Bengal in the north, Uttar Pradesh in the west, and Jharkhand in the south. It also shares an international border with Nepal. Bihar's proximity to the ports of Haldia and Kolkata of West Bengal, accessibility to markets of eastern India, and supply of raw materials from other adjoining states make it economically prosperous, since the ages. The state has experienced higher growth than the national average, the growth rate of Bihar in 2019–20 is 10%; the state per capita GDP of Bihar at the current price was Rs 47,541 and at the constant price is Rs. 33,629 (Government of Bihar 2020).

Bihar, the erstwhile Magadha Empire, has a glorious past of giving the first political map of India, free medical facilities, and world-class universities like—Nalanda

University and Vikramshila University. Bihar also produced the first President in the modern times of India. The state has also been acknowledged as the best administered state of India in the fifties. However, as it faces the challenges in contemporary times—with a lack in the creation of job opportunities to absorb the workforce and the considerably lower ranks in the Sustainable Development Goals Index (SDGs) for the target year 2030 and Global Multidimensional Index 2018, the state presents an alarming concern which is in dire need for initiatives.

The unique location and convenient accessibility to neighboring states and cities of Nepal provide an edge for trade; the vast agrarian society of Bihar is the largest producer of vegetables in the country and engaged almost 74% of the population for their livelihood. At present, Bihar has a chain of industrial areas and industrial parks.

The mismatch in job market of Bihar is a major challenge for youths in the state. The challenge of unemployment forced them to migrate to other states for a better opportunity. The state has undertaken the initiatives to improve the skill level to match the market requirements. The Government of Bihar with the support of the Union Government is imparting skill education through 16 departments under the umbrella of the Skill Development Mission. When witnessed holistically, health, agriculture, and labor—all the welfare departments should work together. Although, bringing them together is a tough task. Further, the aim continues to give the young population better wages (Kline and Moretti 2013).

The state of Bihar is mainly an Agriculture-driven state, the opportunities for employment and empowerment can be generated only through the investment in the state. Bihar is an agricultural state with 80–85% of employment in Agriculture. The natural inclination in such a case is of course agriculture or food processing value added. Yet the big Indian companies like Parle, Godrej, Amul, etc. which have such a capability are not making use of such potential.

Interestingly, the state of Bihar lacks in the areas like basic infrastructure, electricity, transport, connectivity to ports, etc. Hirschman's examination shows that in unbalanced growth in many countries, balanced growth was not a process but a goal to be achieved through a series of imbalances over a period of time (Hirschman 1958). Development can only take place by un-balancing the economy. This is possible by either investing in Social Overhead Capital (SOC) or in Directly Productive Activities (DPA). The former is a suitable option for the state, since the Big-ticket investments require basic services without which primary, secondary, and tertiary productive activities cannot function. In fact, the investment in SOC is advocated not because of its permits but rather because it invites DPA to come in.

The Government of India is working in line with the aforementioned, with its schemes focusing on the infrastructure, 24-h electricity, etc. This shall be beneficial for agriculture in the advanced stages, thus welcoming the food processing investments. However, there is a dire need for value addition in the state of Bihar. Although the private sector investments have started approaching the state, the increase of such investment largely depends on the return on such investments. Thus, it is still a far cry in terms of implementation.

The bigger challenge is to improve the ratio of girls to boys in primary, secondary, and tertiary education. There were large disparities in the economic, social, and schooling opportunities available to men and women in Bihar. Gender disparities begin early in life, with female infants having a lower chance of survival than male infants, owing largely to the parental neglect of the female infants. The initiatives like Mukhyamantri Balika Cycle Yojana launched in the year 2006 have delivered significant results. The state government has reported that not only has the percentage of female candidates increased in schools but the total number of female dropouts has also been reduced. As of now, the amount spent on the scheme by the government is 174.36 crores. Such schemes are important as they increase the literacy levels of the females, and the girls are able to complete class 10 or 12; further, this leads to the decline in the fertility rates. The mobility aspect dealt with such a scheme has delivered noticeable results.

The initiative of the Bihar government of 50% reservation of women in the rural governance of Bihar was another path-breaking initiative to encourage women to join mainstream politics. Such a pro-women decision of the reservation of women in panchayats helped in women empowerment. However, there is scope for improvement as the women mukhiyas are already challenged in terms of digital literacy, minimal education, and zero political experience. In the interviews, it was found that most women leaders ended up as proxies the real decision-makers—their husbands or the men in their families. It was found that most women leaders still look to men for their guidance on critical issues and their proxy roles have almost become institutionalized in some places. These areas can be worked upon to bring out the desired results of the policy initiatives.

References

Abrahamsen R (2004) The power of partnerships in global governance. Third World Q 1453–1467

Adhikari A, Bhatia K (2010) Can we bank on banks? Econ Polit Wkly

ADB (2010) Information and communication technology for development ADB experiences. Asian Development Bank, Manila, Philippines

Afridi F, Mukhopadhyay A, Sahoo S (2016) Female labour force participation and child education in India: the evidemce from national rural employment gurantee scheme. IZA J Lab Dev 1–27

Ahluwalia MS (2020) Economic performance of states in post-reforms period. Econ Polit Wkly 1637–1648

Ajaeevka, TISS (2017) State consultative meeting on labour migration from Bihar. Centre for Development Practice and Research, Patna. Available on: https://tiss.edu/uploads/files/Policy_Brief__State_Consultative_Meeting_on_Labour_Migration__from_Bihar.pdf

Alkire S, Foster J (2011) Counting and multidimensional poverty measurement. J Public Econ, 476-487

Alkire S, Roche JM (2013) How multidimensional poverty went down: dynamics and comparisons. University of Oxford, Oxford. Available on: https://www.ophi.org.uk/wp-content/uploads/How-Multidimensional-Poverty-went-down.pdf

Allen C, Metternicht G, Wiedmann T (2019) Prioritising SDG targets: assessing baselines, gaps and interlinkages. Sustain Sci 14:421–438. Available on: https://link.springer.com/article/10.1007%2Fs11625-018-0596-8

Anand R, Tulin V, Kumar N (2014) India: defining and explaining inclusive growth and poverty reduction. IMF working paper

Bank W (2019) The state of social safety nets: 2018. The World Bank Group, Washington, DC
Benassy-Quere A, Bourguignon F, Dercon S, Estache A, Gunning JW, Kanbur R et al (2010) The millennium development goals: an assessment. In: Spence RK (ed) Equity and growth in a globalizing world. The World Bank, Washington, DC, pp 17–40. Available on: https://openknowledge.worldbank.org/bitstream/handle/10986/2458/548910PUB0EPI11C10Dislosed061312010.pdf?sequence=1&isAllowed=y
Bhalla SS (2013) The unimportance of NREGA. Indian Express, July 24. Available on: https://indianexpress.com/article/opinion/columns/the-unimportance-of-nrega/
Bloom DE, Canning D, Jamison DT (2004) Helath, wealth and wellfare. Financ Dev 10–15
Bue MC, Klasen S (2013) Identifying synergies and complementarities between MDGs: results from cluster analysis. Soc Indic Res 647–670. Availble on: https://link.springer.com/article/10.1007%2Fs11205-013-0294-y
Census (2001) Census of India 2001. Registrar general and census commissioner, New Delhi, India. Available on: https://censusindia.gov.in/census.website/
Census (2011) Census of India 2011. Registrar General & Census Commissioner, India, New Delhi. Available on: https://censusindia.gov.in/2011Census/pes/Pesreport.pdf
Chakraborti D, Rahman MM, Chatterjee A, Das D, Das B, Nayak B, Roy MM (2016) Fate of over 480 million inhabitants living in arsenic and fluoride endemic Indian districts: magnitude, health, socio-economic effects and mitigation approaches. J Trace Elem Med Biol Organ Soc Miner Trace Elem (GMS) 33–45. Available on: https://doi.org/10.1016/j.jtemb.2016.05.001
Deaton A (2003) Health, inequality, and economic development. J Econ Lit 113–158
Desai S, Vashishtha PS, Joshi OS (2015) Mahatma Gandhi national rural employment guarantee act: a catalyst for rural transformation. National Council of Applied Economic Research, New Delhi
Desjardins R, Schuller T, Feinstein L (2006) Understanding the social outcomes of learning. In: Desjardins R, Schuller T, Feinstein L (eds) Measuring the effects of education on health and civic/social engagement. OECD, pp 11–15
Dreze J (2021) Where is the strategy for dealing with learning loss during Covid? The Indian Express, 17 Sept. Available on: https://indianexpress.com/article/opinion/columns/where-is-the-strategy-for-dealing-with-learning-loss-during-covid-7511344/
Dreze J, Dev N, Khera R (2008) Corruption in NREGA: myth and reality. The Hindu, Jan 22
Drèze J, Gupta P, Khera R, Pimenta I (2019) India's public distribution system after the food security act. Econ Polit Wkly 36–47
Engström RE, Destouni G, Howells M, Ramaswamy V, Rogner H, Bazilian M (2019) Cross-scale water and land impacts of local climate and energy policy—a local Swedish analysis of selected SDG interactions. Sustainability 242–247. https://doi.org/10.3390/su11071847
Esteves T, Rao K, Sinha B, Roy SS, Rao B, Jha S et al (2013) Agricultural and livelihood vulnerability reduction through the MGNREGA. Econ Polit Wkly 94–103. Available on: http://www.jstor.org/stable/24477901
FAO (2020) The state of food security and nutrition in the world. Food and Agriculture Organization, Rome
Ferreira FH, Lugo MA (2013) Multidimensional poverty analysis: looking for a middle ground. World Bank Res Obs 220–235. Available on: https://doi.org/10.1093/wbro/lks013
Fukuda-Parr S (2004) Millennium development goals: why they matter. Glob Gov 10(4):395–402
Global Hunger Index Report 2019 (2019) Cocern Worldwide, Dublin
Government of Bihar (2011) The Bihar state free and compulsory education of children rules, 2011. Human Resource Development Department, Government of Bihar, Patna. Available on: https://www.education.gov.in/sites/upload_files/mhrd/files/upload_document/bihar-rte-rules_english%20(1).pdf
Government of Bihar (2016) Economic survey 2015–16 of Government of Bihar. Ministry of Finance, Governmnet of Bihar, Patna. Available on: https://www.adriindia.org/centre/report_details/economic-survey-2015-16-of-government-of-bihar

Government of Bihar (2020) Economic Survey 2019-2020. Ministry of Finance. Government of Bihar, Patna
Haines A, Smith KR, Anderson D, Epstein PR, McMichael AJ, Roberts I et al (2007) Policies for accelerating access to clean energy, improving. The Lancet 370:1264–1281
Hazarika I (2012) India at the crossroads of millennium development goals 4 and 5. Asia Pac J Public Health 450–463
Hirschman AO (1958) Strategy of economic development. Yale University Press, Yale
Hulme D, Wilkinson R (2012) The millennium development goals and beyond: global development after 2015. Routledge
Ibisch PL, Hoffmann MT, Kreft S, Pe'er G, Kati V, Biber-Freudenberger L et al (2016) A global map of roadless areas and their conservation status. Science 354(6318):1423–1427. Available on: https://doi.org/10.1126/science.aaf7166
IEA (2021) India energy outlook 2021: world energy outlook special report. International Energy Agency, Paris. Available on: https://iea.blob.core.windows.net/assets/1de6d91e-e23f-4e02-b1fb-51fdd6283b22/India_Energy_Outlook_2021.pdf
ISFR (2013) India state of forest report 2013. Ministry of Environment Forest and Climate Change. Government of India, Forest Survey of India, Dehradun
ILO (2018) Women and men in informal economy: a statistical picture, 3rd edn. International Labour Organization, Geneva. Available on: https://www.ilo.org/wcmsp5/groups/public/---dgreports/---dcomm/documents/publication/wcms_626831.pdf
ILO (2021) Young persons not in employment and education (NEET) in India: 2000–2019. International Labour Organization, New Delhi. Available on: https://www.ilo.org/wcmsp5/groups/public/---asia/---ro-bangkok/---sro-new_delhi/documents/publication/wcms_789243.pdf
ITU (2021) Individuals using internet (from any location by gender and urban/rural location %). International Telecommunication Union, Geneva. Available on: https://www.itu.int/en/ITU-D/Statistics/Pages/stat/default.aspx
Jacob N (2008) The impact of NREGA on rural-urban migration: field survey of Villupuram District, Tamil Nadu. CCS working paper no 202. Centre for Civil Society
Kadiyala S, Barnett T (2004) Econ Polit Wkly 39(19):1888–1892
Kapoor A, Sahoo RL (2004) Agricultutralists and the moneylender: a study of rural borrowing in Bihar and Orissa 1912–1936. In: Proceedings of the Indian history congress. Indian History Congress, pp 491–500. Available on: https://www.jstor.org/stable/44144764
Kaushal KK, Melkani VK (2005) India: achieving the millennium development goals through non-timber forest products. Int Forest Rev 128–134
Khera R, Nayak N (2009) Women workers and perceptions of the national rural employment guarantee act in India. In: Gender pathway out of poverty: rural employment. FAO-IFAD-ILO, Rome, pp 1–19
Kline P, Moretti E (2013) People, places, and public policy: some simple welfare economics of local economic development programs. Working paper 19659. NBER. Available on: https://www.nber.org/system/files/working_papers/w19659/w19659.pdf
Krishnan A (2018) Behind India's leap in ease of doing business. The Hindu, 11 Nov. Available on: https://www.thehindu.com/business/Economy/behind-indias-leap-in-ease-of-doing-business/article25469900.ece
Kroll C, Warchold A, Pradhan P (2019) Sustainable development goals (SDGs): are we successful in turning trade-offs into synergies? Palgrave Commun 5. Available on: https://doi.org/10.1057/s41599-019-0335-5
Kumar A, Valecha N, Jain T, Dash AP (2007) Burden of malaria in India: retrospective and prospective view. Am J Trop Med Hygiene 69–78
Lahiry S (2014) India's millennium development goals. World Aff J Int Issues 18(3):140–149
Loiwal M (2020) India Today, 31 Mar
Manning R (2009) Using indicators to encourage development: lesson from millennium development goals. DIIS. Danish Institute for International Studies, Copenhagen

Mathy S, Blanchard O (2016) Proposal for a poverty-adaptation-mitigation window within the green climate fund. Clim Policy 752–767. Available on: https://doi.org/10.1080/14693062.2015.1050348

Mattos FB, Dasgupta S (2017) MGNREGA, paid work and women's empowerment. International Labour Organization, Geneva. Available on: https://www.ilo.org/wcmsp5/groups/public/---ed_emp/documents/publication/wcms_613735.pdf

McMichael AJ, Friel S, Nyong A, Corvalan C (2008) Global environmental change and health: impacts, inequalities, and the health sector. BMJ 191–194

Mehrotra S, Sinha S (2017) Explaining falling female employment during a high growth period. Econ Polit Wkly 54–62

Meier GM (1976) Leading issues in economic growth. Oxford University Press, New York

Ministry of Education (2021) Initiatives by the school education sector in 2020–21 for continuing teaching and learning. Minstry of Education Government of India, New Delhi. Available on: https://dsel.education.gov.in/sites/default/files/update/covid_initiatives.pdf

Ministry of Rural Development (2011) Ministry of Rural Development, Government of India, New Delhi

mint (2016) Bihar extends 35% reservation for women in all govt jobs. mint, Jan 2016. Available on: https://www.livemint.com/Politics/jNlMwoB5KuaHUZX767EbeI/Bihar-extends-35-reservation-for-women-in-all-govt-jobs.html

Mohanty SK (2012) Multiple deprivations and maternal care in India. Int Perspect Sex Reprod Health 38(1):6–14

MOSPI (2021) Millennium development goals India Country Report-204-15. Ministry of Statistics and Programme Implementation. Government of India, Central Statistics Office, New Delhi

MOSPI (2017a) Millennium development goals—final country report. Central Statistics Office, Government of India, New Delhi. Available on: http://mospi.nic.in/sites/default/files/publication_reports/MDG_Final_Country_report_of_India_27nov17.pdf

MOSPI (2017b) Millennium development goals—final country report of India. Ministry of Statistics and Programme Implementation, Government of India, New Delhi

MOSPI (2017c) Millennium development goals India country final report. Ministry of Statistics and Programme Implementation. Government of India, Central Statistics Office, New Delhi

MOSPI (2021) Sustainable development goals national indicators framework: progress report 2021. Ministry of Statistics and Programme Implementation, Government of India, New Delhi. Available on: http://mospi.nic.in/sites/default/files/publication_reports/SDG-NIF-Progress2021_March%2031.pdf

Moyo S, Wright JA, Ndamba J, Gundry S (2004) Realising the maximum health benefits from water quality improvements in the home: a case from Zaka district, Zimbabwe. Phys Chem Earth 1295–1299

Nelson RR (1956) A theory of the low-level equilibrium trap in underdeveloped economies. Am Econ Rev 894–908. Available on: http://www.jstor.org/stable/1811910

Nerini FF, Julia Tomei LS, Bisaga I, Parikh P, Black M, Borrion A et al (2017) Mapping synergies and trade-offs between energy and the sustainable development goals. Nat Energy 10–15. Available on: https://www.nature.com/articles/s41560-017-0036-5

Neve J-ED, Sachs JD (2020) The SDGs and human well-being: a global analysis of synergies, trade-offs, and regional differences. Sci Rep 1–12. Available on: https://www.nature.com/articles/s41598-020-71916-9

Nilsson M, Griggs D, Visbeck M (2016) Policy: map the interactions between sustainable development goals. Nature 320–322. Avaoilable on: https://www.nature.com/articles/s41599-019-0335-5#ref-CR15

NIPUN BHARAT (2021) National initiative for proficiency in reading with understanding and numeracy: guideline for implementation. Department of School Education & Literacy, Ministry of Education, Government of India, New Delhi. Available on: https://www.education.gov.in/sites/upload_files/mhrd/files/nipun_bharat_eng1.pdf

NITI Ayog (2019) SDG India. Index & Dashboard 2019-20. Government of India, New Delhi. Retrieved from: https://www.niti.gov.in/sites/default/files/SDG-India-Index-2.0_27-Dec.pdf
NSSO (2014) Participation of women in specified: activities along with domestic duties. NSS 68th round. Ministry of Statistics and Programme Implementation, Government of India, New Delhi
NSSO (2015) NSS report no. 566: status of education and vocational training in India. Ministry of Statistics & Programme Implementation, New Delhi. Available on: https://mospi.gov.in/documents/213904/301563//nss_report_no_566_21sep151602577933947.pdf/a5443bf8-4a37-ae08-f4f5-85f125bd1a2d
NSSO (2020) Annual report: periodic labor force survey (PLFS). Ministry of Statistics and Programme Implementation, Government of India, New Delhi
Oommen OV, Laladhas KP (2016) Sustainable management and use of terrestrial ecosystem. In: RIS, India and sustainable development goals: the way forward. Research and Information System for Developing Countries (RIS), New Delhi, pp 181–194
Osborn D, Cutter A, Ullah F (2015) Universal sustainable development goals: understanding the transformational challenges for developed countries. United Nations, New York. Available on: https://sustainabledevelopment.un.org/content/documents/1684SF_-_SDG_Universality_Report_-_May_2015.pdf
Pangariya A, Mukim M (2014) A comprehensive analysis of poverty in India. Asian Dev Rev 1–52. Available on: https://documents1.worldbank.org/curated/en/847861468257945118/pdf/946460v10JRN0B0O0900170adev0a000021.pdf
Piot P, Greener R, Russell S (2007) Squaring the circle: AIDS, poverty, and human development. PLoS Med 4(10):e314, 1571–1575
Planning Commission (1985) Seventh five year plan: 1985–1990. Planning Commission of India, New Delhi
Poalnki P (2012) How UP and Bihar are skewing India's poverty estimates. First Post, 20 Mar. Available on: https://www.firstpost.com/business/economy/how-up-and-bihar-are-skewing-indias-poverty-estimates-250226.html
PLFS (2020) Annual report 2018-2019. Ministry of Statistics and Programme Implementation, New Delhi. Available on: http://mospi.nic.in/sites/default/files/publication_reports/Annual_Report_PLFS_2018_19_HL.pdf
Poku N, Whitman J (2011) The millennium development goals: challenges, prospects and opportunities. Third World Q 3–8
Pradhan P, Luís Costa DR, Kropp JP (2017) A systematic study of sustainable development goal (SDG) interactions. Earth's Future 1169–1179. Available on: https://doi.org/10.1002/2017EF000632
Purvis K (2016) 12 ways environment and development sectors can collaborate to meet the SDGs. The Guardian, 31 Oct
Radkar A, Parasuraman S (2007) Maternal deaths in India: an exploration. Econ Polit Wkly 3259–3263
Rao CS (2007) Bridging the Telecom Divide. Econ Polit Wkly, 209-211
Ravallion M, Chen S (2003) Measuring pro-poor growth. Econ Lett 93–99
Ravallion M, Datt G, Walle DV (1991) Quantifying absolute poverty in the developing world. Rev Income Wealth 345–361
Ravindranath NH, Murthy IK (2021) Mitigation co-benefits of carbon sequestration from MGNREGS in India. PLoS ONE
RBI (2021) Table 10: per capita net state domestic product—state-wise (at constant prices). Reserve Bank of India, New Delhi. Available on: https://www.rbi.org.in/scripts/PublicationsView.aspx?id=20415
Registrar General of India (2011) Censul 2011. Registrar General of India, Government of India, New Delhi
Robert, Fioramonti, Kubiszewski I (2016) The UN sustainable development goals and the dynamics of well-being. Front Ecol Environ 59. Available on: https://doi.org/10.1002/fee.1231

Roberts K (1980) Price-independent welfare prescriptions. J Public Econ 13:277–297. Available on: https://www.sciencedirect.com/science/article/abs/pii/0047272786900071
Samuelson PA, Swamy S (1974) Invariant economic index numbers and canonical duality: survey and synthesis. Am Econ Rev 566–593. Available on: http://www.jstor.org/stable/1813311
Scherer L, Behrens P, Koning A, Heijungs R, Sprecher B, Tukker A (2018) Trade-offs between social and environmental sustainable development goals. Environ Sci Policy 65–72. Available on: https://doi.org/10.1016/j.envsci.2018.10.002
SECC (2011) SECC 2011 at a glance. Ministry of Rural Development, Government of India, New Delhi. Available on: https://secc.gov.in/getSeccDataSummaryNationalReport.htm
Sen A (1976) Poverty: an ordinal approach to measurement. Econometrica 219–231
Sen A (1979) Equality of what? The Tanner lecture on human development, 22 May, pp 195–220
Sen A (1999a) Development as freedom. Knopf, New York
Sen A (1999b) Development as freedom: an India perspective. Indian J Ind Relat 157–169
Sen A (2001) Many faces of gender inequality. Frontline. Available on: https://www.sas.upenn.edu/~dludden/MANY%20FACES%20OF%20GENDER%20INEQUALITY.htm
Sen A (2005) The three R's of reform. Econ Polit Wkly 1971–1974. Available on: http://www.jstor.org/stable/4416610
Sen A, Dreze J (2011) Putting growth in its place: it has to be but a means to development, not an end in itself. Outlook, 14 Nov. Available on: https://www.outlookindia.com/magazine/story/putting-growth-in-its-place/278843
Shah M (2007) Employment guarantee, civil society and Indian democracy. Econ Polit Wkly 43–51. Available on: http://www.jstor.org/stable/40276832
Sharma R (2018) Health and economic growth: evidence from dynamic panel data of 143 years. PLoS ONE. https://doi.org/10.1371/journal.pone.0204940
Sharma AN, Saluja MR, Sharma A (2016) Macro economic impact of social protection progrrames in India. Econ Polit Wkly 121–126
Singh M (2012) PM's opening remarks at the meeting of PM's national council on skill development. Prime Minister Office, New Delhi. Available on: https://archivepmo.nic.in/drmanmohansingh/content_print.php?nodeid=1126&nodetype=2
Singh UK (2020, September 20) Even after 15 Years of Nitish Kumar, Bihar Remains an Anarchic State and Poor. National Herald, pp. Available on: https://www.nationalheraldindia.com/opinion/even-after-15-years-of-nitish-kumar-bihar-remains-an-anarchic-state-and-poor
Singh UK (2021) India has a serious hunger problem and it needs urgent policy intervention. The WIRE, 30 Oct. Available on: https://thewire.in/food/india-has-a-serious-hunger-problem-and-it-needs-urgent-policy-intervention
Singh NK, Stern N (2013) The new Bihar. HarperCollins
Smith P, Olesen JE (2010) Synergies between the mitigation of, and adaptation to, climate change in agriculture. Agric Sci 543–552. Available on: https://doi.org/10.1017/S0021859610000341
Tewary A (2020) Tejashwi Yadav's apology for 'mistakes' of RJD regime fails to impress Bihar's ruling NDA. The Hindu, 3 July. Available on: https://www.thehindu.com/news/national/other-states/tejashwis-apology-for-mistakes-of-rjd-regime-fails-to-impress-bihars-ruling-nda/article31982878.ece
Thapar R (1966) A history of India. Penguin Books, London
Thomas D, Frankenberg E (2002) Health, nutrition and prosperity: a microeconomic perspective. Bull World Health Organ 106–113
TNN (2019, February 12) Bihar's Urban Tele-density Second Best in Country. Times of India, pp. Available on: https://timesofindia.indiatimes.com/city/patna/bihars-urban-teledensity-second-best-in-country/articleshow/67947741.cms
UN (1972a) Report of the United Nations conference on the human environment. United Nations, New York. Available on: https://undocs.org/en/A/CONF.48/14/Rev.1
UN (1972b) United Nations conference on the human environment, Stockholm, 5–16 June 1972. United Nations, New York

UN (1987) Report of the world commission on environment and development: our common future. United Nations, New York. Available on: https://sustainabledevelopment.un.org/content/documents/5987our-common-future.pdf
UN (1992a) Report of the United Nations conference on environmemt and development, vol I. United Nations, New York. Available on: https://undocs.org/en/A/CONF.151/26/Rev.1
UN (1992b) United Nations conference on environment & development: agenda 21. United Nations, New York
UN (1992c) United Nations conference on environment and development. United Nations, New York
UN (2002) Report of the world summit on sustainable development. United Nations, New York. Available on: https://undocs.org/pdf?symbol=en/A/Conf.199/20
UN (2012a) Resolution adopted by the General Assembly on 27 July 2012. United Nations, New York. Available on: https://www.un.org/ga/search/view_doc.asp?symbol=A/RES/66/288&Lang=E
UN (2012b) Resolution adopted by the General Assembly on 27 July 2012. Agenda No. A/RES/66/288. United Nations, New York
UN (2013) Open working group of the General Assembly on sustainable development goals. Agenda No. A/67/L.48/Rev.1. United Nations, New York
UN (2015a) Resolution adopted by the General Assembly on 25 September 2015. A/RES/70/1. United Nations, New York
UN (2015b) Resolution adopted by the General Assembly on 25 September 2015. Agenda No. A/70/L.1. United Nations, New York. Available on: https://www.un.org/ga/search/view_doc.asp?symbol=A/RES/70/1&Lang=E
UN (2015c) When we stand together, there is no limit to what we can achieve, Ban tells UN Assembly. United Nations, New York. https://www.un.org/sustainabledevelopment/blog/tag/secretary-general-ban-ki-moon/
UN (2016a) MDG monitor: MDG 6: combat HIV/AIDS, malaria and other major diseases. United Nations, New York
UN (2016b) Report of the inter-agency and expert group on sustainable development goal indicators. Economic and Social Council, United Nations, New York. Available on: https://undocs.org/E/CN.3/2016/2/Rev.1
UN (2017) The sustainable development goals report: 2017. United Nations, New York. Available on: https://unstats.un.org/sdgs/files/report/2017/thesustainabledevelopmentgoalsreport2017.pdf
UN (2020a) Policy brief: the impact of COVID 19 on children. United Nations, New York. Available on: https://unsdg.un.org/sites/default/files/2020-04/160420_Covid_Children_Policy_Brief.pdf
UN (2020b) The sustainable development goals report: 2019. United Nations, New York. Available on: https://unstats.un.org/sdgs/report/2019/The-Sustainable-Development-Goals-Report-2019.pdf
UN (2020c) United Nations in India. United Nations, New York. Available on: https://in.one.un.org/page/sustainable-development-goals/gender-equality-sdg-5/
UN (2021) The sustainable development goals report: 2021. United Nations, New York. Available on: https://unstats.un.org/sdgs/report/2021/
UNAIDS (2021) Global HIV & AIDS statistics—fact sheet. Retrieved from: https://www.unaids.org/en/resources/fact-sheet
UNDP (1997) Human development report 1997. United Nations Development Programme, New York
UNDP (2015) Building the post-2015 development agenda: 2014 annual report. United Nations Development Programme, New York. Available on: https://www.undp.org/sites/g/files/zskgke326/files/publications/UNDP_Building%20the%20Post-2015%20Development%20Agenda_Final_2015.pdf
UNDP (2018) Human development report 2018. United Nations Development Programme and Oxford University Press, New York

UNDP (2019) Global multidimensional poverty index 2019: illuminating inequalities. United Nations Development Programme and Oxford Poverty and Human Development Initiative, New York. Available on: http://hdr.undp.org/sites/default/files/mpi_2019_publication.pdf

UNICEF (2016) A snapshot of Bihar elementary education: district-wise statistical report. UNICEF BIHAR, Patna. Available on: https://www.bepcssa.in/en/A%20Snapshot%20of%20Bihar%20Elementary%20Education1.pdf

UNICEF (2017) A world free from child poverty: a guide to the tasks to achieve the vision. United Nations Children's Fund, New York

UNICEF (2020a) Maternal mortality in 2000: estimate developed by UNICEF and UNFPA. United Nations Children's Fund, New York

UNICEF (2020b) Universal child benefits: policy issues and options. UNICEF, New York. Available on: https://www.unicef.org/media/72916/file/UCB-ODI-UNICEF-Report-2020.pdf

UNICEF India (2016) Children in India: India's growth over the last two decades has contributed phenomenally to global human development. United Nations Children's Fund, New Delhi

Wheebox (2021) India skills report 2021. Wheebox, Gurugram. Available on: https://indiaeducationforum.org/pdf/ISR-2021.pdf

WHO (1990) Maternal mortality in 2000: estimate developed by UNICEF and UNFPA. World Health Organization, Geneva

WHO (1991) Maternal mortality: a global factbook. World Health Organization, Geneva

WHO (2015) Global tuberculosis report 2015. World Health Organization, Geneva. Retrieve From: https://apps.who.int/iris/bitstream/handle/10665/191102/9789241565059_eng.pdf?sequence=1&isAllowed=y

WHO (2019) Maternal mortality. World Health Organization, Geneva

WHO (2020) Global report on WASH in health care facilities. World Health Organization, Geneva

World Bank (1993) World Bank development report 1993: investing in health. The World Bank Group & Oxford University Press, Washington, DC

World Bank (2008) World development report 2008: agriculture for development. The World Bank, Washington, DC. Available on: https://doi.org/10.1596/978-0-8213-6807-7

World Bank (2009) Doing business in India 2009. The World Bank, Washington, DC. Available on: https://openknowledge.worldbank.org/handle/10986/28088#:~:text=Doing%20business%20in%20India%202009%20is%20the%20first%20country%20specific,cities%2C%20and%20documents%20their%20progress

World Bank (2013) Information and communication technologies: Results profile. The World Bank Group, Washington, DC

World Bank (2017) New World Bank project to give impetus to India's skill's agenda—to provide over 8 mn youth with market-relevant training. The World Bank Group, Washington, DC

World Bank (2020) Global economic prospects, June 2020. The World Bank, Washington, DC. Available on: https://openknowledge.worldbank.org/handle/10986/33748

World Bank (2021) World Bank indicators. The World Bank, Washington, DC. Available on: https://data.worldbank.org/indicator/SP.DYN.IMRT.IN

Worldometers (2020) Coronavirus cases. Worldometers. Available on: https://www.worldometers.info/coronavirus/coronavirus-cases/?fbclid=IwAR1rXCB1NYf48HuNqhHZDPNSUySwz8PpWwqaGRsE-3PijhQfc03axd2FPcI

Zhao Z, Cai M, Wang F, Winkler JA, Connor T, Chung MG, Zhang J, Yang H, Xu Z, Tang Y, Ouyang Z, Zhang H, Liu J (2021) Synergies and tradeoffs among sustainable development goals across boundaries in a metacoupled world. Sci Total Environ 1–18. Available on: https://doi.org/10.1016/j.scitotenv.2020.141749

GPSR Compliance

The European Union's (EU) General Product Safety Regulation (GPSR) is a set of rules that requires consumer products to be safe and our obligations to ensure this.

If you have any concerns about our products, you can contact us on ProductSafety@springernature.com

In case Publisher is established outside the EU, the EU authorized representative is:

Springer Nature Customer Service Center GmbH
Europaplatz 3
69115 Heidelberg, Germany

Batch number: 10837251

Printed by Printforce, the Netherlands